THE SOUTH STAFFORDSHIRE COALFIELD

THE SOUTH STAFFORDSHIRE COALFIELD

NIGEL A. CHAPMAN

First published 2005

Reprinted in 2008 by
The History Press
The Mill, Brimscombe Port,
Stroud, Gloucestershire, GL5 2QG
www.thehistorypress.co.uk

Reprinted 2013

© Nigel Chapman, 2005

The right of Nigel Chapman to be identified as the Author
of this work has been asserted in accordance with the
Copyrights, Designs and Patents Act 1988.

All rights reserved. No part of this book may be reprinted
or reproduced or utilised in any form or by any electronic,
mechanical or other means, now known or hereafter invented,
including photocopying and recording, or in any information
storage or retrieval system, without the permission in writing
from the Publishers.

British Library Cataloguing in Publication Data.
A catalogue record for this book is available from the British Library.

ISBN 978 0 7524 3102 4

Typesetting and origination by
Tempus Publishing Limited.
Printed in Great Britain.

Contents

	Acknowledgements	6
	To Win the Coal	6
	Introduction	7
one	The South Staffordshire Coalfield	9
two	Black Country Collieries	19
three	The Big Four Collieries	53
four	New British Iron Co.	81
five	Cannock Chase Collieries	91

Acknowledgements

Over the years as the opportunities have arisen, many photographs have been collected or copied as an archive of Black Country mining. It was hoped that one day they could be made available as a book to record something of a vanished industry which did much to create the West Midlands as it is today.

Photographs have been borrowed and copied from many individuals, and to these I tender my grateful thanks. They include: Ray Shill, Alan Hill, Jeff Parkes, Peter Knowles, Alan Hickling, John Selway and Simon Chapman, Claire Hetherington and the staff at Sandwell Archives and Reference Libraries, the staff past and present in the Science & Technology Department of the Birmingham Central Libraries for information and help over many years, and Lisa Smith and the staff of Cannock Libraries, Reference Department.

To Win the Coal

South Staffordshire is favoured with the finest bed of coal that has yet been discovered and by far the greater portion of this magnificent seam of Thick or Ten Yard coal has been raised by small engines geared off by small pinions and large crown wheels to shafts bearing niche rings into which are wound heavy flat rivet or wood blocked chains.

In too many instances may now be seen one of these engines standing in the midst of from four to eight shafts, some over a hundred yards distant, all of which are worked from it.

The cumbrous chains travel along lines of rollers paced upon posts and over pulleys about 4 feet diameter surmounting three ugly legs which compose what is called a pit frame.
(The *Engineer* of 27 October 1871)

Introduction

This description of a typical Black Country colliery of the nineteenth century shows how simple the average colliery of the period was. It consisted of two shafts, sometimes brick lined, of about 7ft diameter sunk to the famed Thick Coal seam of the coalfield. In each shaft was a chain built up of three separate chains composed of long and short links placed close together with wooden wedges hammered into the long links. This formed the standard winding chain of the area. As it was wound over the pulleys and drums it made such a noise that it was known as 'Rattle Chain'. Naturally these collieries soon acquired the name of Rattle Chain pits. On the end of each chain in the shaft would be hung a skip, a flat wagon with the lumps of Thick Coal piled up and held in place by iron hoops. To wind from these shafts a beam engine powered by steam was used which slowly raised skip and chain from one shaft while an empty skip was lowered down the other. At the shaft bottom the Hooker On took the skip off the chain and it went off into the workings in charge of a horse and driver. While at the shaft top a large square flat wooden door on wheels was pushed over the shaft and the loaded skip lowered on to it. Detached from the chain the skip could be taken away to be tipped on the pit bank. Because of the shallow depth of the coal seams, mostly about 100 yards, and the huge thickness of the coal in the Black Country, a very wasteful system of mining developed. Many colliers who saved some money were able to obtain a charter and work a colliery for a landlord such as the Earl of Dudley. They used any equipment available, and the accidents and deaths were just part of normal life in a colliery.

By 1870 the use of cages to carry tubs of coal up and down the shafts was becoming more common and the old beam engines were being replaced by fast horizontal steam winding engines. The two small shafts continued to be sunk with single- or double-deck cages run in these shafts, with wooden or iron guides to keep the cages in place in the shafts. If water was encountered then a shaft was deepened and a 'Sump' formed to collect water while a tank was put into that shaft and wound as part of the normal winding cycle. The Thick Coal seam produced firedamp but no form of artificial ventilation was usually employed because of the fear of spontaneous combustion. The Thick Coal was very prone to catch fire even when in tips or on rare occasions in the holds of ships at sea. It was believed that the iron pyrite in the coal was heated by the friction of moving air and so fired. The only ventilation permitted was from the heat of colliers and horses underground, which helped the air to go up one shaft and cause air to descend the other shaft. In mining terms the shaft with the descending air was known as the Downcast and the ascending air shaft was the Upcast. Should a strong wind develop on the surface then the air flow could change, and it is said that the air flow went in one direction in the summer and the opposite during winter.

I hope this fairly basic description provides readers with some impression of a typical small Black Country Colliery. Should further explanation be required, I would suggest a visit to the Black Country Museum and a tour of their colliery.

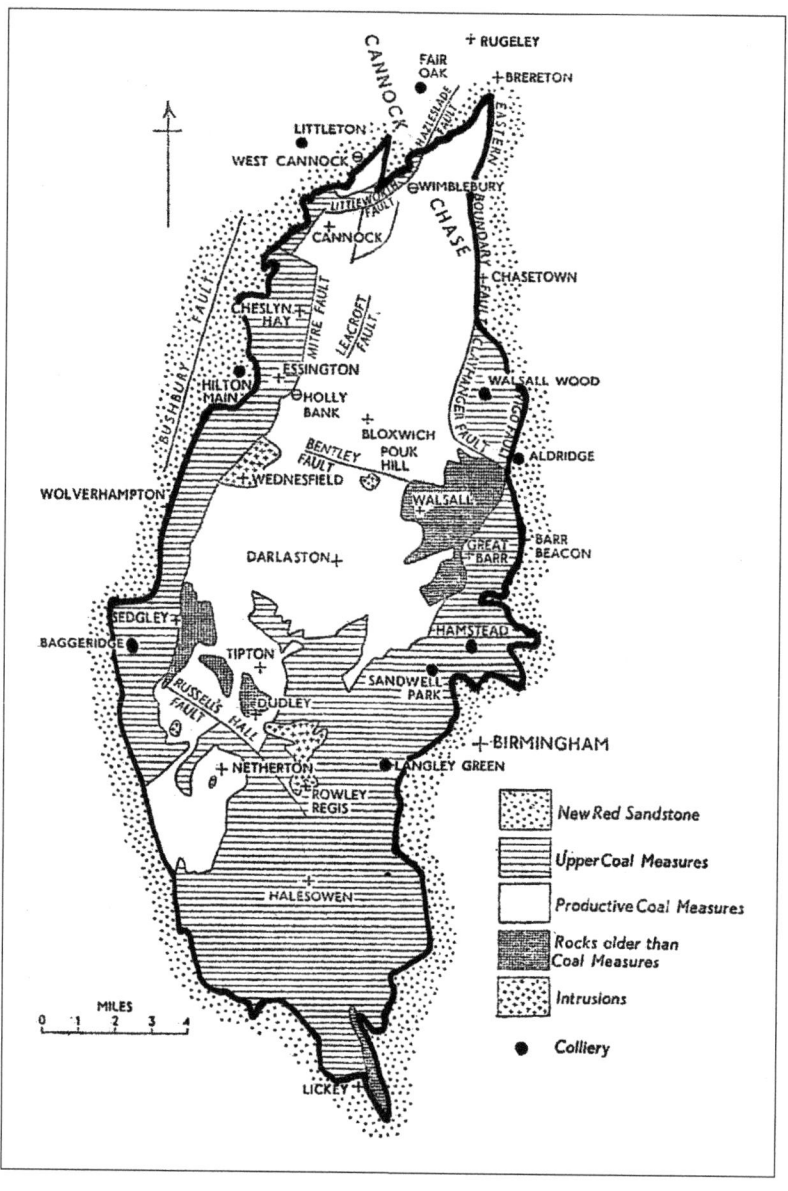

Sketch map of the South Staffordshire Coalfield.

one

The South Staffordshire Coalfield

The Middle coal measures of the Carboniferous Series forming the South Staffordshire coalfield stretch in an unbroken line for fifteen miles from Stourbridge in the south to Brereton in the north. The exposed portion of the coalfield is found at depths from 100 to 600ft below the ground in the shape of a rough triangle with an apex at Brereton, widening out gradually to Essington in the south-east and across to Pelsall in the south-west. This northern portion of the coalfield was separated from the south by a major fault running east and west near Bentley. With a down-throw of over 100m, this fault broke up the famous Thick Coal seam of the Black Country into about fifteen different thin seams, while it pushed the coal seams deeper underground. The coalfield also suffered from a series of large faults which cut the seams into separate areas, while lifting some of these seams of form areas of shallow coal, it meant that for most of the area, seams were deeper underground than in the Black Country. Partly for this reason, the Cannock area developed at later date and required greater expense in sinking to the deeper seams, therefore larger royalties were leased before operations commenced. North of the Bentley fault the coal produced was more suited to domestic rather than industrial purposes, instead of the wide range of uses made of the Thick Coal. From these remarks it become apparent that the Bentley fault was responsible for major differences in the geology of the Northern section which were reflected in the different characteristics of the two parts of the coalfield. Cannock Chase had three major coal seams, the Shallow and Deep, while in the Brownhills area the Brooch seam was much sought after.

Early coal mining was to be found in areas such as Essington, Pelsall, Brownhills, Great Wyrley and Brereton. By 1500 coal was being mined at Brereton and to the south on Cannock Chase, where large areas of tips and hollows remain to this day under the trees planted by the Forestry Commission. Coal mining is recorded in the Essington and Great Wyrley parishes by the seventeenth century, but was of a local nature until the construction of canals in this period gave access to wider markets such has Birmingham. The major development of the Cannock area required the construction of a complex system of railways to allow the sinking of collieries away from the canals. When by the mid-nineteenth century the London & North Western Railway linked the Cannock coalfield to markets far and near, the scene was set for the greatest expansion of mining the area would see. First to see the potential of the area was John R. Mclean, an engineer surveying for the proposed railways. He took a lease of the Hammerwich Colliery in 1854 and laid the foundations of the Cannock Chase Colliery Co. As they developed collieries, raised coal and declared profits others began to follow their lead and sink new collieries around Cannock.

The other pioneer of mining was William Harrison who began the Brownhills Colliery Co. in 1849 by leasing areas of coal from Phineas Hussey of Little Wyrley Hall and gradually worked the seams to the north and east. He was to be involved in the Cannock & Rugeley Colliery Co. and the family were to reopen the Mid Cannock Colliery.

* * *

Early coal mining was concentrated on the Wednesbury to Oldbury area, working the thickest seams of coal at the shallow depth of up to 300ft. Because of the relative ease of access, small pits working areas of several acres and run by working colliers with very little capital became the standard for the Black Country. With the thickest coal seam in the country to work, these basic collieries employed old plant and backward methods which lead to great wastage of coal and colliers' lives. Even as late as the 1960s such small collieries were still to be found in one or two locations in the Black Country.

In areas with coal seams close to the surface, the winding plant would have been a horse gin raising wicker baskets carrying the coal. These were still to be seen into the 1920s with iron tubs instead of wicker baskets for winding on the coalfield.

As the Thick Coal seam was prone to outbreaks of spontaneous combustion, no form of artificial ventilation was permitted, yet naked lights were always used. Natural ventilation powered by the heat of the colliers, horses or the outside temperature changes were used. On occasions explosions of methane gas wreaked havoc in the pit and killed or burnt the colliers. By 1870 ventilating fans were being erected at several collieries but these were never popular. Many pits were ventilated by jets of steam released at the shaft bottom to create a column of rising air and draw the shale air and gasses out of the workings.

A PAIR OF THE EARL OF DUDLEY'S THICK COAL PITS IN THE BLACK COUNTRY

Above: A drawing to set the scene of coal mining in the Black Country, late in the nineteenth century. The Earl of Dudley's Collieries. The industrialisation of the area is well shown in this illustration.

Right: The collieries at Broadwaters were the scene of the first attempt to raise water from the mines by steam power in 1706. The mines had been suffering from water seeping into them from the Broadwaters Lake which had accumulated over collapsed workings. Existing forms of pump had failed to stop the water and the use of a Savery pump was tried but proved to be a complete failure and the mines were drowned and lost.

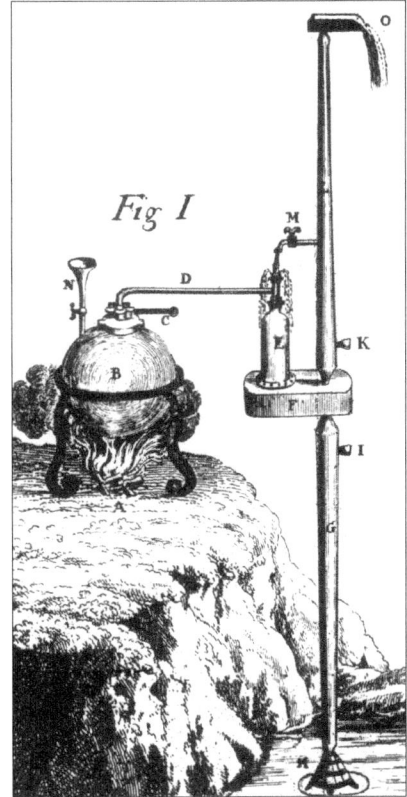

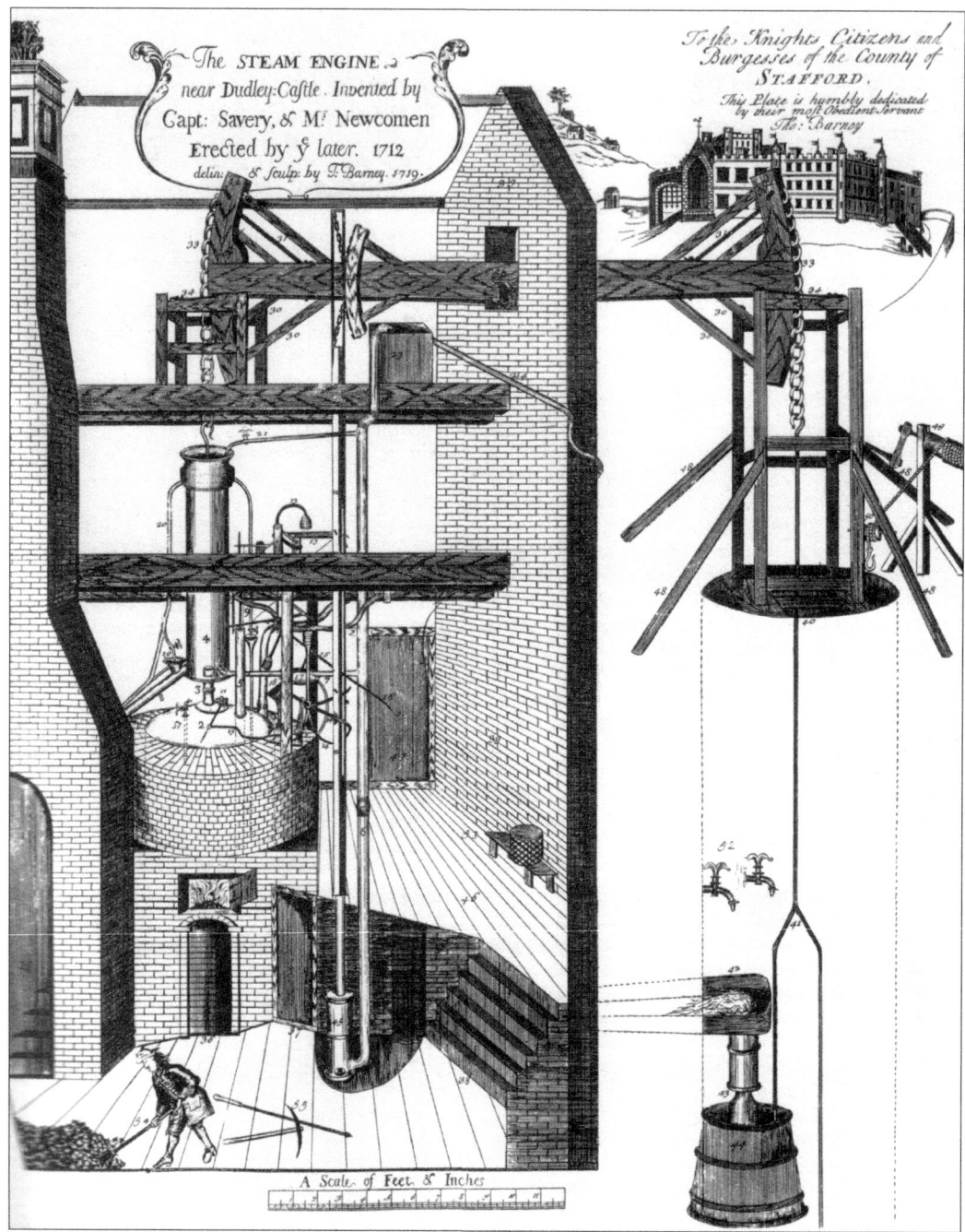

Newcomen Engine. Developed by Thomas Newcomen of Dartmouth, the first working atmospheric engine was erected in 1712 at Coneygre for the Earl of Dudley. This engine was the subject of much interest and was drawn by T. Barney in 1719. The site of this engine is believed to be within the Black Country Museum and a modern replica has been built not far from the original site.

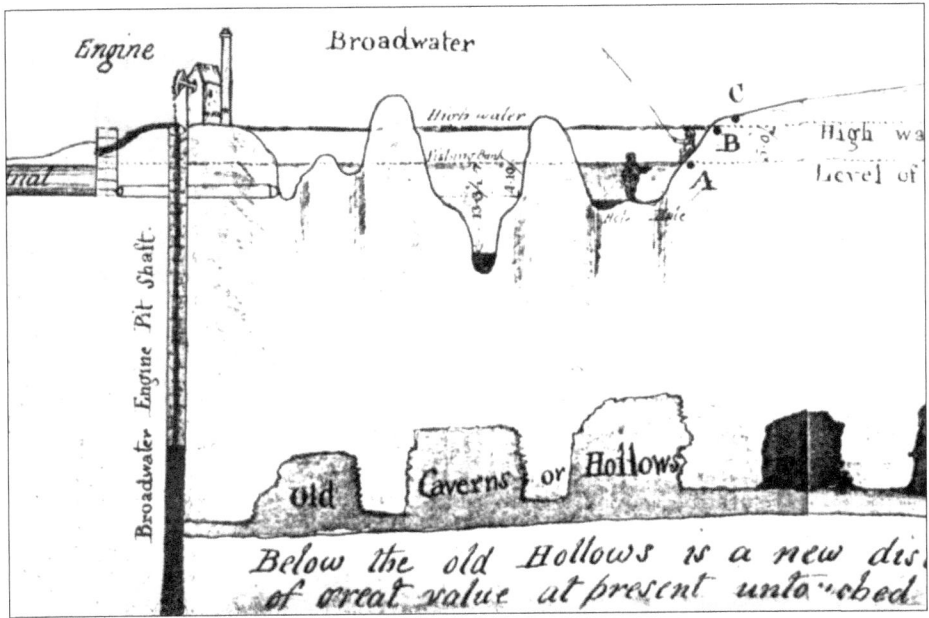

Drawing of the Newcomen pumping engine on a shaft at Broadwaters. Believed to have been the engine of 1712 moved from Coneygre, the engine pumped the water from the flooded mines after the failure of the Savery pump. Draining the mines was so successful that in 1808 Thomas Telford could build his Holyhead Road across the middle of what had formerly been Broadwaters Lake.

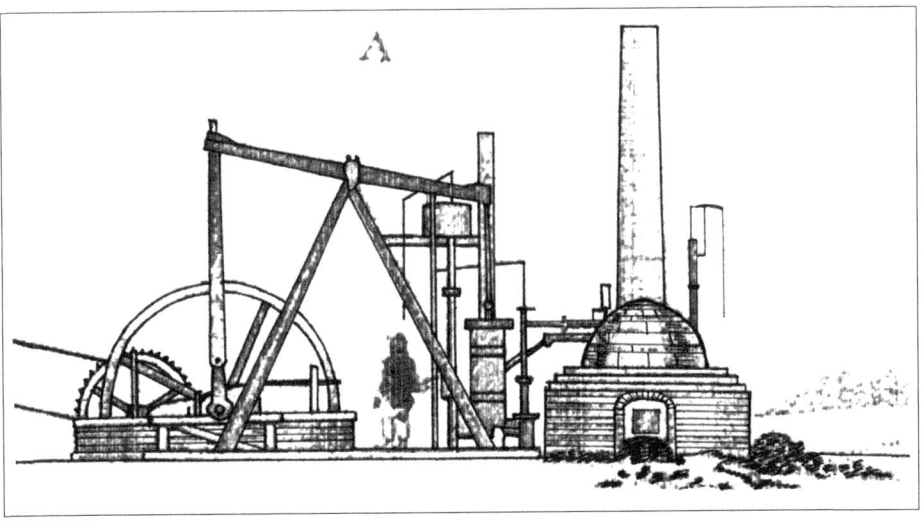

By 1800 the Newcomen engine was to be found winding coal from many of the deeper collieries of the coal field. This drawing shows the basic components with a balloon boiler on the right supplying steam to the cylinder to operate the beam and turn the flywheel, gearwheels and drum on the left. No engine house was erected as the engine would be used at this shaft for a year or two then moved to a new site. The last of these winders operated into the early 1900s.

Close up of the beam winding house at the Shut End Colliery, probably No.17 Pit. The spoked wheels on the left were to collect and guide the Rattle Chain into place on the reel drum. Once common across the coalfield, the last Rattle Chain winder is believed to have been at the Shut End Colliery until 1915.

Opposite above: Fosters Field No.6 Pit Pensnett, a Rattle Chain pit with heavy chain over posts between the beam winder and the pit frame. On the left is the tall brick engine house for the beam winder with boiler plant and water supply. Very basic primitive wooden frames stand over the shafts with two upright posts for guides of the cages in the shafts. This was a colliery from about 1840 standing derelict in the 1920s.

Opposite below: No.7 Pit of the Himley Colliery in around 1900 with Rattle Chain, guide posts and a head frame in the background. To the rear of that is a horse gin with an openwork rope drum. The man with the bowler hat would be the 'Doggie' or Foreman, who was often also the charter-master working the colliery.

One of the Stallings Lane pits with Rattle Chain reels and the simple brick-built house. Note no guards, apart from a low wooden fence – a health and safety nightmare! This was around 1900.

A simple cage on the top of a shaft with a group of colliers. The large wooden frame on top of the cage was used to cover the open shaft when the cage had left the surface. It was lifted by the cage on reaching the surface.

Right: A more substantial cage at a colliery with a larger work force in the Wednesbury area, about 1900.

Below: Pit Bank Wenches said to be at the Wednesbury Blue Fly Colliery, probably about 1890. Women were employed at Black Country collieries to pick coal, move tubs and some served to land tubs at the shaft top. They were never employed underground in this area.

Left: A group of nineteenth-century colliers at Wednesbury.

Below: A drawing of one of the Saltwells Colliery workings about 1850. A number of colliers are busy cutting in various areas of the workings. A skip is being loaded in the left foreground with large pieces of the Thick Coal held in place by iron hoops. The horse leads an empty skip with several iron hoops on it from the shaft ready to be loaded. The only roof support noted consists of a pillar of coal usually known as 'A Man of War'. The railway was composed of plateway sections of 'L'-shaped rail while the wheels were flangeless. Lighting in the pit was by candles placed in suitable positions to illuminate operations for the colliers.

two

Black Country Collieries

By the middle of the nineteenth century the South Staffordshire Coalfield, an area of about 50 square miles, became known as the Black Country. It was said to have a character all of its own, covering the townships of Dudley, Stourbridge, Walsall, Wednesbury, Darlaston, Bilston and Wolverhampton. What made it so unique was the exploitation of the Ten Yard or Thick Coal seam lying at a shallow depth below the surface. Often associated with the coal were seams of ironstone, while near Dudley and Walsall beds of limestone were to be found. Towards Stourbridge two seams of fireclay became of value, leading to the creation of a major brick-making industry in the area. Some of these collieries were raising coal and fireclay, often with the fireclay being the most important. Then the coal was used to heat the kilns to produce fine quality bricks from the fireclay.

The term 'The Black Country' is a sobriquet applied in about 1850 to an area west of Birmingham. It was an unofficial name and the area it covered has no precise definition, but it is generally accepted to be the area above the famed Thick Coal seam of the South Staffordshire and East Worcestershire coalfield. Generally the area was from Bilston in the north to the River Stour and slightly further south, about fourteen miles and from West Bromwich in the east to Kingswinford in the west, about ten miles. Over the 140 square miles of the Black Country, many small industrial villages developed and some were to become substantial towns with manufacturing specialities, such as Wednesbury for iron tubes and Walsall for leather and horse equipment, while Bilston developed a large steel-making industry. Darlaston developed an industry in the production of screws. Cradley Heath became known for the quality of its chain making, Brierley Hill for steel making and iron products, Wordsley for its glass, Halesowen for nails and Lye for galvanising. Smethwick was to develop in Messrs Chance some of the finest glass produced in the country, much of their products going into lighthouses to luminate the sea-lanes of the world. Oldbury was to provide a home for the major tube producers, Accles and Pollock, while the Albright & Wilson Chemical Works also found a home in the town.

A contemporary traveller in the Black Country described the area in the following terms:

> Huge spoil banks of black shale, great mounds of rough slag and heaps of cinders and refuse of all descriptions surround dark and massive smelting furnaces, puddling furnaces, rolling mills and other huge structures for the smelting and manufacture of iron. Flames and smoke are ceaselessly belched forth from their summits, while from their murky recesses break forth

dazzling rays of light emitted by the surface of molten iron through half open furnace doors or millions of brilliant sparks starting from the glowing lumps at each blow of the forge hammer.

Around these cyclopean dens are innumerable coal pits, with their engine houses and their creaking machinery for pumping and winding, large coke hearths with their smouldering heaps, and great stacks of coal and iron ready for transport along the network of canals and railways and tramways that multiply in every direction among them. The peculiarity of the scenery is increased by the occasional appearance of houses, sometimes of large buildings and high chimneys, slanting considerably from the perpendicular in consequence of the sinking of the ground on which they stand, while the same causes produce unnatural looking hollows, filled sometimes with shallow pools of water spreading here and there over the surface. The canals and railways for the same reason require constant watching and frequent repairs and additions to their sides and embankments to keep them up to their original level. If we add to these strange features clouds of dense smoke and black dust, and imagine the perpetual creaking and clanking of chains and machinery, the panting and shrieking of steam engines, the continual thumping of forge hammers and the everlasting bellowing of blast furnaces, we shall complete a picture calculated to astonish.

This was the Black Country in the 1850s, an area in which some of my ancestors toiled.

Opposite above: Sunk in 1834 the Haden Hill Colliery No.1 Pit is shown at work about 1890. Horses wait patently for the carts to be loaded for the journey to local houses. More coal and tubs stand on the railway nearby.

Opposite below: Haden Hill No.1 Pit with chimneys, a couple of head frames and mining clutter looking across the canal towards Black Heath. This was about 1890. The colliery was sunk in 1834 to work the Thick Coal seam and ironstones. It ceased coal winding about 1896 leaving the steam winder idle; this was converted to drive a generator to supply electric power to haulage motors underground and lights on the surface.

Coal being hand-loaded into barges at the Haden Hill basin. These collieries were producing electricity by 1896, permitting lights to be placed on the wharf for the night loading of coal.

A busy screen at the Haden Hill Colliery basin. Coal in tubs from pits Nos 1 and 2 were delivered here to be loaded into barges for transport through the Gorsty Hill tunnel to Birmingham.

Narrow-gauge incline from Haden Hill No.2 Pit down to the canal basin with the chimney of the Waterfall Lane Mines Drainage plant in the background. Constructed around 1865, the incline was operated by an engine driving an endless rope which was continually moving. To the rope were clipped the individual tubs for their journey up or down the incline.

Haden Hill Colliery No.2 Pit was sunk in 1865 to a depth of 263 yards to work the famed Thick Coal seam. Because of the trees and shrubs planted to hide the colliery from the landowner's home at Haden Hill Hall it was also known as 'The Pretty Pit'. After several closures and reworking the colliery finally closed in 1926.

Left: Maintenance work taking place on the boiler chimney at Haden Hill No.2 Pit. On the far left of the picture the third shaft at this colliery can be seen. This was only used to supply good quality water from the sandstones to the boiler plant.

Below: A busy scene at the Rowley Hall Colliery with a steam-operated beam winding engine in the middle and a head frame on each side over the shafts. On the left empty tubs wait to be sent underground. Sunk in 1856, the colliery closed in 1918. During the sinking of the shafts the local Rowley Rag stone was found and a large quarry developed on the surface. This stone was supplied to Councils in the area for road making.

Above: Shaft top at Rowley Hall Colliery with a dog standing in the doorway to the winding house. The shaft is protected by a wooden cover between the two uprights.

Below: From Rowley Hall Colliery a tramway was constructed to the canal and tubs of coal were lowered by steam engine down to the basin. The ropes were carried on sheaves set in the track bed to reduce wear.

Rowley Hall canal basin with barges loaded and empty. On the right-hand wall are two rotary tipplers for loading barges directly from the mine tubs. Electric lights are provided for night work.

An atmospheric wasteland at the Homer Hill Colliery. Sunk in 1867, the colliery was the scene of an explosion which caused the deaths of twelve colliers and injuries to ten others. The Mines Inspector demanded a means of ventilation and the first Guibal Fan on the coalfield was erected here in 1868.

Opposite above: Section of the workings at Homer Hill Colliery in 1871. A major and several small faults, known as 'Troubles' for obvious reasons, are shown. Each shaft was sunk to work different levels of the coal seams.

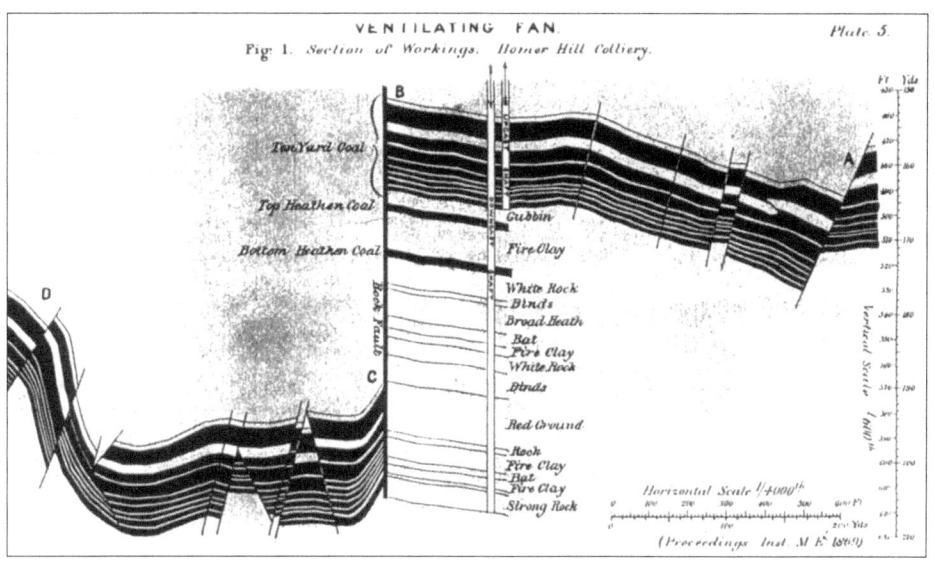

Windmill End Colliery No.3 Pit. Sunk about 1830 this was one of the Earl of Dudley's collieries near the Netherton Canal Tunnel. A large beam pumping engine was in the house to the left. Later used by the Oldhill Mines Drainage Co., the plant was still in situ in 1928. The second house held a Newcomen winding engine of 1830 date. When closed in 1928 this engine was brought by Henry Ford and shipped to his museum in Dearborn, Michigan, where it can still be seen.

Windmill End today. The buildings are now protected as part of the Coal Mining Heritage of the area. They have been restored and are scheduled as ancient monuments.

Himley Colliery No.8 Pit near Dudley. Sunk about 1870 this colliery was operated by the Earl of Dudley's mining department. They developed an unusual steam winding engine with a set of outside drums and flywheel as depicted in the photograph. Within the building was a single-cylinder horizontal steam engine to operate the outside drums. Constructed by the Earl of Dudley's foundry at Castle Mill, these unique winders were only used at the Earl's collieries. Only three photographs are known to exist of this type of winder and the much battered remains of one engine house still exists.

Right: Himley Colliery No.8 Pit near Dudley. Sunk about 1870, this is a further view of this unique form of winding engine with the flywheel and drums outside the engine house. The only remaining example of this type of building at the Himley Colliery No.9 Pit was in a derelict state until bulldozed about 1976. The date stone and the shaped sandstone blocks for the windows can now be seen in the gate house of the Pensnett Trading Estate to the west of Russell's Hall Hospital, Dudley.

Below: In areas of shallow mining the horse gin was much used for winding the coal. During the twentieth century much of the earlier shallow workings were reopened and coal extracted by primitive means often by small groups of colliers. In this example two separate shafts have been sunk and the tubs clipped onto the chain to be lifted up the shaft.

An example of a Gin Pit near Burton Road Dudley about 1900. In this area the Thick Coal seam outcropped, so shallow mines with horse gins were to be seen. Used from the earliest days of mining in the Black Country, they could still be found at shallow mines into the 1920s.

An abandoned colliery still complete with the wooden head frames and wheels over the shafts. When sales of coal were poor, some of these mines would be closed and stand idle for several years. Then either the former charter-master or new ones would clean the engine, repair the workings and coal would be produced again. This was probably in the Oldbury area about 1900.

A derelict beam winding engine and boiler plant probably around 1900. In the background is a pair of abandoned blast furnaces. This was probably in the Dudley area.

In the Wednesbury area, Thick Coal was found just under the surface, being exploited from medieval times. At Kings Hill some quarries were opened in the Thick Coal seam to extract the workable areas of coal. With the tools placed to provide a scale, a good impression of the thickness of the coal seam is given.

The Thick Coal suffered from spontaneous combustion, which was said to have occurred because of the iron pyrites in the coal. Here in Wednesbury in about 1900 the 'Wild Fire', as it was known, destroyed the road and was threatening adjacent houses. During 1890 a Council watchman who was protecting passers by from the fires fell into the burning void and was killed. Even as late as 1936 areas of coal were still burning in Old Park. During the 1970s the waste tip at the New Hawne Colliery was on fire, this too caused by the effects of spontaneous combustion many years after the colliery closed.

A scene at a coal wharf in Wednesbury in about 1890, with heaps of coal tipped on the edge of the canal ready to be loaded into carts. Many coal merchant's yards across Staffordshire would have looked similar, with plenty of hard work for the shovelling team.

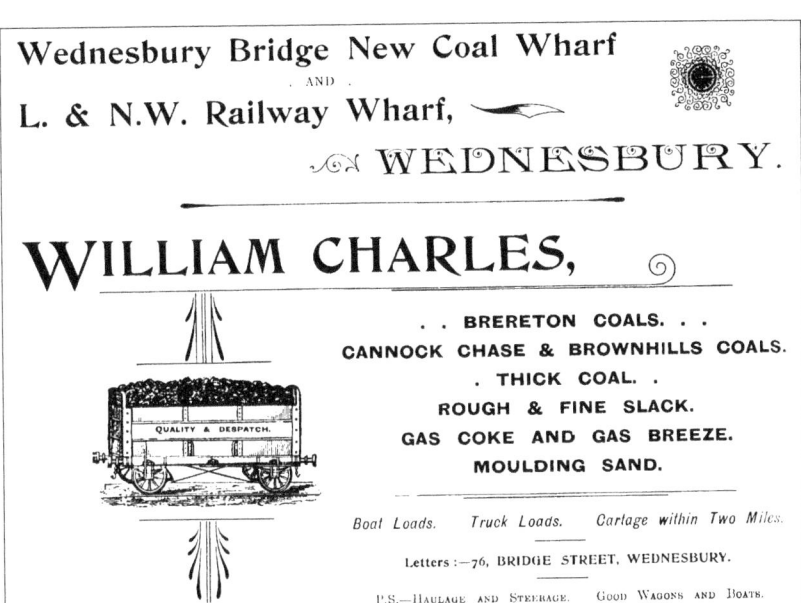

Nineteenth-century advertisement placed in the local trade gazetteer showing the variety of coals stocked and the locations they came from.

John Bagnall & Sons was a well-known local firm having originated from the Shropshire Coalfield in the seventeenth century. They operated in 1873 the furnaces at Gold's Green, and Capponfields, and also the Toll End, Lea Brook, Imperial and Gold's Hill Ironworks. Coal and ironstone was raised from their Gold's Green, Cophall, Bentley, Deepmoor, Capponfield, James Bridge, Crescent, Bradley, Groveland and Pump House Collieries.

Above: John Bagnall & Sons Lea Brook Ironworks was on the side of the canal for ease of transport of raw materials to the works and manufactured goods to overseas countries. The Thick Coal seam was composed of several bands of coal, some providing furnace coal, others were used for forging or Smith' works, while some bands supplied house coal.

Left: Buffery pumping engine, Netherton. Still retaining its wooden beam, this fine example of a Watt steam pumping engine of about 1790 worked until the end of the nineteenth century. In later years it was used by the South Staffordshire Mines Drainage Commission to reduce the levels of water in the local mines. Eventually a new pumping plant was built at Waterfall Lane, Oldhill.

Blue Fly Colliery at Wednesbury about 1890, showing the landsale wharf with a line of coal carts being loaded with large coal. This was the most prized coal, selling at a higher price, and was loaded carefully by hand. A number of women appear to be engaged in loading the carts.

The colliery near Salop Street in Dudley with the tower of a windmill in the background. The waste from the colliery spread across the ground stops anything else from growing. This was a small pit, probably selling its coal directly to local coal merchants.

Above: Witley Colliery, Halesowen, sunk in 1872, was unusual in having a large diameter main shaft with two cages operating in it. The brick winding house in the background held a large horizontal steam winder which by 1872 was replacing the beam winder still used at older collieries. The directors of the company were from North Wales or Lancashire and created their ideas of a colliery. This colliery was to produce coal and fireclay; both were used to produce a fine quality firebrick for lining the furnaces of the Midlands.

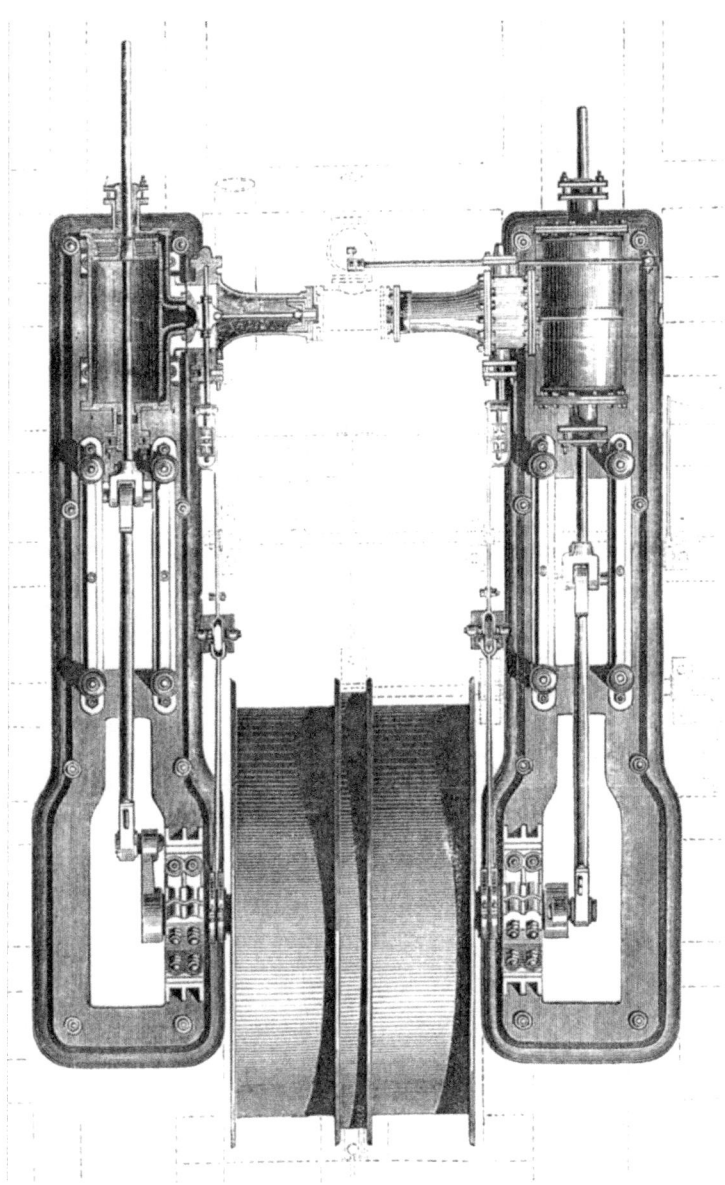

Above: Grace Mary Colliery. Sinking of the No.2 colliery was completed in 1870 and what was probably the first horizontal steam winder in the Black Country was set to work. This was an occasion for the Dudley Mine Agents Association to have a visit to the colliery and tour the workings, with a lunch and some speech making; the occasion was reported in the mining press of the day.

Opposite below: The first Grace Mary Colliery was sunk about 1842 near the Birmingham Canal and linked by tramway to a basin on the canal side. When in 1870 Samuel Minton decided to sink a new colliery higher to the hillside, to work new areas of coal seams, he had a twin-cylinder horizontal winder manufactured. It was built by Messrs Withinshaw of Wiggin Street, Birmingham, being probably the first horizontal winder on the coal field. The winder was also the first application of round wire ropes to a colliery in the coal field instead of the use of chains. The shafts were sunk to a depth of 284 yards, having cut the Thick Coal of 34ft thickness at a depth of 258 yards.

Jagger Carts in Wednesbury. Poor families wanting to purchase coal found the full load of a cart too expensive, so these smaller carts were developed to deliver portions of a cart load. All the wheels must have made hard work for the poor horse.

Oldnall Colliery near Halesowen. Sunk in 1874 by James Fisher to a depth of 191 yards, the Thick Coal and fireclays work under lands belonging to the Oldswinford Hospital. Closed in 1887 the colliery was reopened in 1899 by Messrs Mobberley & Perry. They made improvements to the plant of the colliery, operating it until closure in 1944. The tramway on the left was constructed in 1919 for the nearby Beech Tree Colliery, so that the output could be lowered by an incline to their Hayes Works.

From some of the bands which made up the Thick Coal, coke could be made for blast furnace use. Often burnt in the open in large heaps, the coal produced huge clouds of smoke (it was the Black Country!). In later years the Glede Oven was developed for the coking of coal but without any attempt to utilise the by-products burnt out of the coal. This picture shows a typical smoky row of Glede Ovens probably in the Oldbury area.

An underground scene with a collier cutting under the coal while timber props have been set to prevent the coal falling on him. Having cut deeply under the coal with a pick, it was hoped the removal of the timber would bring down the hanging coal.

A view of a coal face with colliers cutting under the hanging coal. The collier on the right is charging a shot hole with explosive to bring down the coal once colliers and timber had been removed.

Two colliers with picks (known as Pikes and therefore Pikemen), cutting round the corner of a pillar. The coal in large lumps had been loaded up into a tub standing on the right. In the nineteenth century a collier worked in a coal mine or colliery while a miner worked in an ironstone mine. However, all mines were locally known as pits.

Three colliers are building a 'Cog' of horizontal timber baulks tightly filled with shale and small coal or 'Slack'. This construction was designed to support the roof and would withstand heavy weights as the roof collapsed. Note the use of a wicker basket by the collier on the right.

Above: Two colliers are depicted in the act of setting a prop. In colliers' terms they are 'Setting a tree'. The thin piece of wood held on top of the tree by the second collier was called a lid. Its purpose was to ease the tree into place. When the prop came to be extracted, the lid helped to ease the work.

Right: Three colliers are shown setting a timber prop in a 'Side of Work', as these huge underground quarries were known; the picture conveys the massive size of these quarries in the Thick Coal.

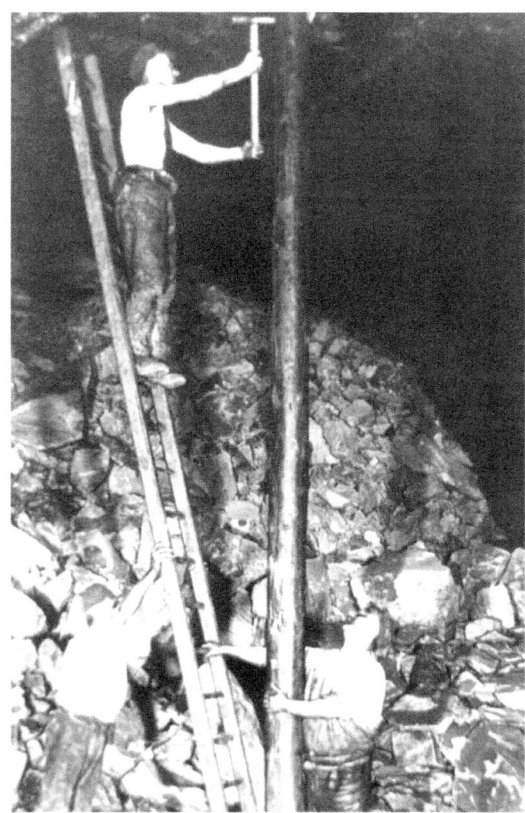

Left: Standing on a board chained to a couple of ladders, the collier cuts the top of the Thick Coal seam. Once the coal made a noise or moved the collier would be down the ladder and gone, if he was lucky! This type of work was considered the most dangerous, being undertaken by the most experienced and therefore the highest paid workers in the colliery. Note the thickness of this superb coal seam which made the Black Country what it once was. Most of these underground pictures were the work of W.H. Hughes, the manager of the Earl of Dudley's collieries in the late nineteenth century. Mr Hughes produced some of the best underground pictures in existence at the time and won prizes for them in various countries.

Below: Two colliers working in a 'Place' with a rotary hand-operated drilling machine. The light colour suggests they are driving a gate-road in rock requiring the use of explosives because of its hardness. Developed to speed work underground, hand drills were much used in the late nineteenth century. Later compressed air drills were introduced to improve the output but saw little use at the collieries of the area until well into the twentieth century.

Having cut the coal and loaded it into a tub, children, men, horses or compressed air machines moved the tubs to the shaft bottom for the journey up to daylight. The Haulage man is securing a set of tubs to a moving wire rope by tightening a clip. This set appears to have come out of a branch heading on the left and is being attached to the main haulage rope in the pit. The chalked numbers on the tub identify the colliers who had filled it; this was the means by which they would be paid.

A bend on a main haulage system with a row of wheels set in the curve to get the rope round the corner. Much time and effort has gone into the support of the roof with the use of steel girders across quite a large tunnel. On the right side are two bare wires used by the haulage men to signal to the engine house. Making the wires touch rang a bell for stop, start and so on, giving instructions for the haulage driver.

Taken during one of the periodic coal strikes, families are busy picking or digging for pieces of coal for the fires at home. During lengthy strikes, picking over the tip of the local colliery became one of the necessities of life with colliery officials and the local police to chase and arrest people for theft. On occasions the holes in the tips collapsed causing several deaths from suffocation.

Coal picking in the Coal Strike, 1912. This orderly scene shows the coal loading wharf of the Corbyn's Hall Collieries with a standard gauge wagon waiting to be loaded from the tubs on the higher level.

Cradley Park Colliery on the hillside above Park Lane, Halesowen. Production of coal commenced in 1867 with Joseph King & Co. as the mine owners. Never a large colliery, it gave employment to about 150 men until it closed on 20 January 1933.

Winding house and a timber frame of the Cuba Pit near Dudley. This was a small colliery which worked from early in the twentieth century until the 1940s. The timber head frame over the shaft appears to have required support from a couple of steel girders.

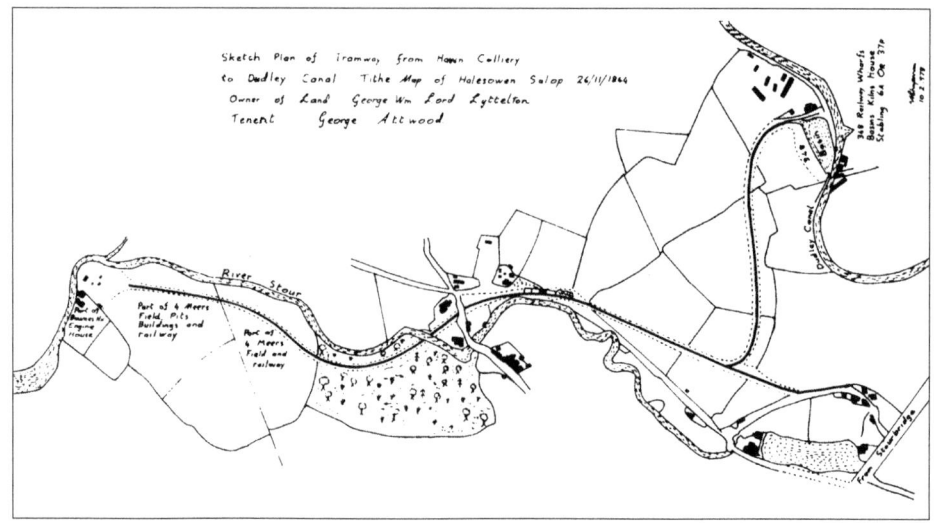

When in 1844 the Title Map of Halesowen was drawn up, the only colliery shown was at The Hawn. Being away from the Dudley No.2 canal a tramway was constructed and operated by horses to take the coal to the newly created Halesowen Docks. Later much used by the local ironworks into the early 1900s, the canal basin is still used today.

Coombs Wood Colliery, sunk in 1908 by Noah Hingley & Co. to supply their ironworks at Oldhill with coal. As the existing Hingley collieries in the Oldhill area closed the colliers were transferred to Coombs Wood and output improved. Under the National Coal Board the colliery was licensed as a private mine, continuing to operate until the rising cost of pumping the mine water forcing closure in 1948. The buildings were taken over for metal processing and remained until they were demolished in the late 1960s.

Messrs H.S. Pitt & Co. reopened many of the old workings in the area of Windmill End at Netherton around 1900 and developed a number of collieries on the hillsides above the Dudley No.2 canal. Under new Coal Mines Acts the formation of mines' rescue teams was demanded of the larger companies and a Mines' Rescue Station was established in Trindle Road in Dudley. H.S. Pitt's No.2 team is depicted at a practice rescue session at the new station in Dudley.

Pennant Hill Colliery. Reopened by H.S. Pitt & Co. around 1900, the colliery had once been worked in the nineteenth century as the Scotwell Colliery. It was to be the scene of an explosion of firedamp on 22 October 1915, causing the deaths of five colliers. The cage was blown out of the shaft and its wrecked remains form the subject of the photograph. A young boy was a survivor of this event and appears in the inset.

Left: Pennant Hill Colliery, showing a rescue team about to descend the damaged shaft using the Bowk after the cage had been destroyed, on 22 October 1915. This was a dangerous task; they would face foul air, gas and the possibility of the shaft walls collapsing, yet miners were easily found for the work.

Below: Amblecote Colliery No.2 Pit, a small colliery working into the 1960s near Brierley Hill. The Downcast steel head frame towers over the corrugated iron surface buildings. In the foreground is a simple lorry loading point.

Amblecote Colliery No.2 Pit. Still using steam power in 1960, the colliery retained much of the atmosphere of a late nineteenth-century Black Country Colliery.

Amblecote Colliery No.2 Pit. The steam winding plant complete with boiler, ornate square chimney and sawmill. This is a view of the plant when it closed during 1962.

Above: Amblecote Colliery No.2 Pit. The steel head frames stand on the mound of waste raised from the shafts. This illustrates why the area around the shaft was known as the 'Pit Bank'.

Left: Amblecote Colliery No.12 Pit. The wooden head frame was still standing in 1960. Worked by John Hall & Co. to supply coal for their brick making works, the colliery was one of the last to work in the Black Country.

Racecourse Colliery No.1 Pit. Using the No.126 Pit of the nineteenth-century Coneygree Colliery, the Black Country Museum Mining Group has created a typical small Black Country pit. The items in the left foreground are parts awaiting restoration of a steam winder from Amblecote Colliery No.12 Pit.

Racecourse Colliery No.1 Pit. With the original steam winder from the Amblecote Colliery No.12 Pit available, the winding house at the Black Country Museum was modelled closely on the house at the No.12 Pit. Restored to steam the winder had been supplied second-hand from Devon to the Amblecote Colliery about 1910.

Twin Pits Colliery, Tividale. Probably sunk into one of the collieries of the former Tividale Colliery, it was worked from the 1840s to the 1860s by Round Brothers of Hange Ironworks. The Twin Pits Colliery Co. was said to have been operated as a tax swindle. The winding house converted into a workshop was in existence into the 1980s.

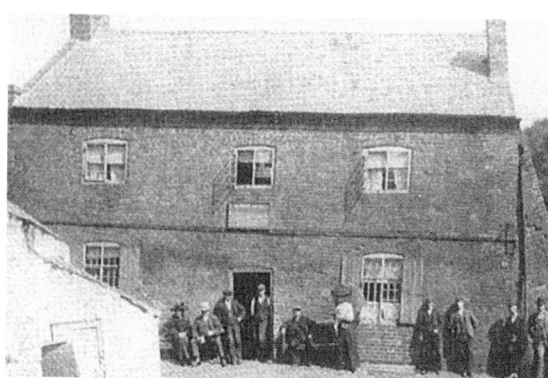

Once a farmhouse of eighteenth-century date, this building was converted into a more profitable trade, selling beer to the colliers at the Himley Colliery. Known as the Glynne Arms, it was undermined by the Himley Colliery working the Thick Coal seam at a shallow depth. In more recent times the public house has been renamed the Crooked House for fairly obvious reasons and still sells beer to passing visitors curious to see this example of mining subsidence.

Netherton Colliery No.10 Pit. In around 1910 a local press photographer took this superb picture of the 'Pit's Company' or the collier's employed at the pit, against the backdrop of the timber head frames over the two shafts. The Netherton Colliery had been working under various owners since at least 1840 and by 1910 it must have been nearly devoid of coal. Most of the men show their occupation by holding candles.

The High Street, Bilston, was the scene of some sewer pipe laying around 1900 and during these operations a coal seam was cut. This was soon turned to profit and a small coal working developed in the street while normal life carried on around it. The hand-operated winding gear, usually known as a 'Jack Roll', stands over the shallow depth shaft.

three

The Big Four Collieries

By the 1860s very little virgin coal could be found within the existing boundaries of the Black Country, so beginning with Sandwell Park collieries over the known boundaries were sunk to greater depths of over 1,000ft to find new reserves of coal. These sinkings required huge investment of capital, therefore to make these collieries viable much larger royalties of land were leased and worked. Each of these four major pits were operated by limited liability companies with equipment, ventilation and methods of mining either imported from other coal mining areas or developed to suit the situation. All four collieries were comparable with the most modern collieries of the period, being very different in most respects from the small collieries normally associated with the Black Country. They were to work the same coal seams as the smaller pits, therefore suffering from the same geological problems, only on a much larger scale. In terms of output they were to produce in a month, what the smaller pits produced in a year and to employ proportionally more colliers. These major pits were to last until the end of mining in the coalfield in the 1970s.

Sandwell Park Colliery

During 1870 Henry Johnson, a mining engineer of Dudley, proposed and eventually commenced the sinking of a trial shaft over the eastern boundary of the coalfield towards Birmingham. This trial, after costing around £20,000, proved the Thick Coal seam to be of 24ft thickness at a depth of 418 yards during April 1874. Two further shafts were to be sunk over the next few years to develop into the first major modern colliery in the Black Country. Producing about 300,000 tons of coal annually, the colliery worked into the 1900s. Further reserves of coal were located to the north but separated by geological faults. From 1897 to 1908 a new colliery known as the Jubilee was sunk and constructed in the Sandwell Valley coming into production in 1909. Coal in tubs was transported to the old Sandwell Park site for processing before shipment either by rail or canal to the works of the area and further afield.

Taken over by the Warwickshire Coal Co. in 1934, the Jubilee Colliery was extensively rebuilt prior to the Second World War. Taken over by the National Coal Board in 1947, the colliery finally closed in September 1960.

Above: Advertised for sale at the Poynton Colliery Cheshire in 1870, this 45in beam pumping engine was re-erected at the Sandwell Park Colliery in 1871. It was used to pump water from the original trial shaft and remained on the colliery until scrapped sometime in the 1950s.

Left: The newly erected steel lattice headgear designed by H.W. Hughes for the sinking of the Jubilee Colliery; it stands over the site of the Downcast shaft.

Right: Sinking the Downcast shaft Jubilee Colliery about 1900. A Bowk of waste rock from the shaft is just being tipped into a wooden tub on the Pit Top.

Below: Jubilee Colliery, Sandwell Valley, taken from the south probably in the 1920s.

Jubilee Colliery, Sandwell Valley, showing the alterations done by the Warwickshire Coal Co. in the 1930s. Downcast shaft is on the right.

Jubilee Colliery, Sandwell Valley. Another 1930s view of the Warwickshire Coal Co.'s alterations. Upcast shaft top and new steel headgear.

Sandwell Park Colliery became the processing plant in the 1930s for coal raised from the Jubilee Colliery. In this 1930s view the new canal boat loading plant is shown. Today this structure forms a major part of the few remains of these collieries.

The canal boat loading plant as it stands today beside the Birmingham Canal.

A rural scene across the cornfields looking towards the tip and chimney of Jubilee Colliery in the 1960s. The second waste tip belonged to Hamstead Colliery further to the east. Today the M5 runs across the middle of this scene, and much of the rural character has changed to a more urban setting.

Believed to be a view of the old Sandwell Park Colliery in the 1950s, the colliery could have been extensively modernised by the National Coal Board, but was it?

Hamstead Colliery

The success of the sinking of the Sandwell Park Colliery to the east of the boundary fault inspired other mining engineers to sink shafts further east in the hope of finding seams of coal. The sinking of the Hamstead shafts commenced in 1875, cutting the Thick Coal seam of 24ft thick at a depth of 615 yards in 1880. Sinking the No.2 shaft and developing the gate roads underground took until 1882. During 1883 57,617 tons were raised. Output by 1892 had risen to 325,000 tons per annum. Hamstead was to suffer a series of accidents, the colliery was abandoned during November 1898 and the shafts sealed to smother a major underground fire. Reopened in 1900 the colliery was to be the scene of a major disaster on 4 March 1908 when a fire in the Downcast shaft caused the deaths of twenty-four colliers and one rescuer. Eventually reopened, the colliery was to be taken over in 1947 by the National Coal Board who spent several millions modernising the surface buildings and installing new equipment. Further fires in the 1960s reduced the output while large areas of coal reserves were sealed off. Closure came in 1964.

Pit Top area about 1900 with the lattice wrought iron headgear built by the Horseley Iron Co. for £1,105. To the right stands the Upcast shaft timber head frame and winding house.

Probably taken in the 1920s, the Downcast shaft headgear has acquired a tall derrick for lifting the pulleys. Over the Upcast shaft now stands a steel girder built headgear. In the pit yard are several vehicles including two early lorries.

Reservoir in the foreground with the chimneys and headgears in the rear. This photograph is dated March 1908, when the disaster was occurring.

Typical postcard put out for a mining disaster, 4 March 1908. This example gives a good view of the colliery at this date. (Compare with the photograph of 1936 of page 63.).

March 1908. With both shafts full of smoke and heat, it was decided to pull air from the Downcast shaft forcing fresh air down the Upcast shaft. By this means rescue teams could be sent underground in fresh air. The round building on the left is a Guibal Ventilating fan which sucked air out of the shaft on the right. To enable it to draw from the Downcast shaft a new air tunnel across the Pit Top was being dug. It would then have been bricked over and tarred to prevent leakages of air.

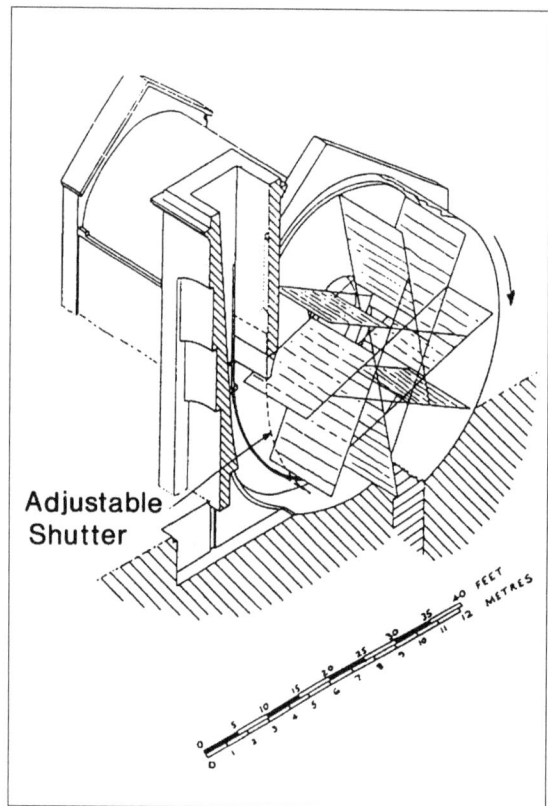

Left: Cut away drawing of a Guibal mine ventilating to show the wooden fan within its housing. It was capable of pulling air out of a mine only.

Below: March 1908. This view shows the new airway as it reached the Downcast shaft. The building left in the background is one side of the multi-deck caging gear. The nearside caging gear has already been demolished for the new airway. This photograph and the preceding one show what was done to overcome a major colliery disaster in a very short period of time.

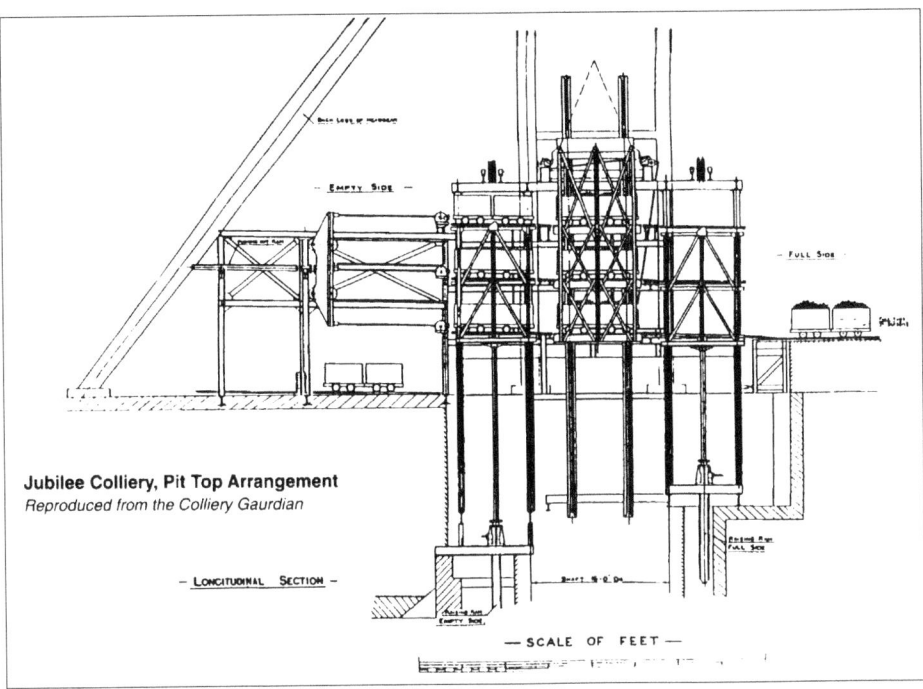

Above: Drawing of caging gear. Hamstead Downcast shaft top looked like this before the disaster, but to assist in the construction of the new airway most of the structures were rapidly demolished as shown in the previous photographs of March 1908. As a direct result of the problems encountered at Hamstead in getting fresh air and rescue teams underground, in 1911 the Coal Mines Acts required the mine fans to be reversible so that air could be pushed into a mine. This requirement still applies today.

Below: View in 1936 of the modernised surface buildings with the new steam winding plant with the electric generation plant. On the right is a large waste tip, once a common sight at a colliery.

Collieries in the Black Country suffered from sudden collapses of the workings caused by the enormous pressures developed because of the huge size of underground workings. Known in the coalfield as 'Bumps' these underground earthquakes as they could be termed destroyed everything in a localised area. Made more ferocious by the increased depth of the workings at Hamstead, they were the subject of much research. This picture illustrates the destructive effects of a 'Bump'.

Baggeridge Colliery

Following an extensive and successful programme of borings in the woods near Himley Hall, the sinking of a 17ft-diameter shaft was commenced during February 1899. This shaft, after meeting an underground river of 60,000 gallons per minute, was completed during July 1902 when at a depth of 556 yards the Thick Coal of 24ft thickness was cut. Sinking of the second shaft cut the Thick Coal seam at a depth of 556 yards on 28 January 1907. Because of the difficulties coal production did not commence until 1912.

Output during 1948 was 422,970 tons from a work force of 1515 men. By 1967 the output had fallen to 121,500 tons. The colliery was closed in March 1968, when it was the last colliery in the Black Country.

Baggeridge Colliery, taken during the 1930s. The coal winding shaft on the right with its steam winding house in the background and a large boiler plant on the left.

Baggeridge Colliery in the 1930s with the large boiler plant in the middle of the picture.

The screening plant, stones and shale were hand picked from the moving belts and loaded into the tubs in the middle. These jobs were done by the older or injured men with a number of boys being trained by them.

The railway sidings with a large overhead loading plant in the background. Coal from the colliery was transported to many works in the Black Country and as far as London.

Aerial view of the colliery about 1950. Clouds of white show that the colliery still used steam winding.

The pit yard in the 1960s with a large number of tubs waiting to be scrapped. In the background stands a lorry loading plant.

Baggeridge Colliery in the 1960s with some alterations to the buildings. In the background stand two chimneys at the brickworks, the only part of this industrial scene still in existence. The colliery site is now part of a country park, while the brickworks once an offshoot of the colliery sends its products all over the country.

Hilton Main Colliery

Mining in the area of Essington had been established by 1851 when the Census noted twenty-seven men were employed. Further development had raised this figure to 188 by 1871 and in 1881 the figure had reached 239 men working in the coal mines.

Of the new mines, Essington Wood Colliery was operated in 1863 by Samuel Mills owner of the Darlaston Iron Co. with an ironworks located at Darlaston Green. Following his death the company was to become the Darlaston Coal and Iron Co. Ltd in 1877 and eventually sold the ironworks in 1882 so that they could concentrate on developing the Essington Wood Colliery. Great difficulties were experienced in the coal trade of this period and this was reflected in the collapse of the Darlaston Coal & Iron Co. Ltd. In 1895 a new company with the same directors was floated under the name of the Holly Bank Colliery Co. and the colliery became Holly Bank. With a new manager, John Charles Forrest, the company was able to overcome their problems and profits were soon being made. Mining to the west in their royalty of 5,000 acres encountered a major fault which threatened to cut off large areas of coal seams. To test the fault a drift was driven across the fault in the 8ft coal seam but encountered a further major fault cutting off all the coal seams of the area. After much geological study it was decided to sink an underground shaft of 14ft diameter to locate the coal seams again.

During 1909–11 the sinking of this shaft was undertaken to a depth of 329 yards and the coal seams found again. New workings were developed from the bottom of this shaft and coal extracted back to the Holly Bank shafts for winding. This system was a useful expedient for a number of years but the extending haulage system plus the complex ventilation arrangements

Holly Bank Colliery Pit No.15. Note the very widely spaced timber legs of this head frame. This would suggest a large diameter shaft with two cages operating in it. Also the untidy lengths of wire guides hanging from the head frame. These were used to keep the cages apart in the shaft, but should have been cut to length instead of being left as depicted.

lead to the consideration of the sinking from the surface of a new shaft. The First World War stopped all new developments until in 1920 the sinking of the new shaft of 18ft diameter was undertaken in Hilton Park. Sinking continued to a depth of 630 yards cutting the Eight Foot coal seam in the process. Workings were developed and a ventilation system established with the Holly Bank Colliery workings making this shaft, a Downcast. Steam winding was used during the sinking but the colliery was one of the first in South Staffordshire to be all-electric. Mining continued until the company liquated in 1930 and was eventually taken over by the Hilton Main and Holly Bank Collieries Ltd in 1932. At this date the colliery was said to have an annual output of 508,380 tons of coal. The new company decided to improve the output and in late 1934 the sinking of a second shaft of 16ft diameter was commenced at Hilton Main. Once this was completed the Hilton Main Colliery was complete and the Holly Bank mines gradually run down and closed. With an output of 2,000 tons of coal per day this colliery was to be taken over by the National Coal Board in 1947 and modernised with new surface plant and buildings plus underground coal cutting machines. During 1967 four seams of coal were being worked, the Bottom Robins of 4ft 6ins thick producing 400 tons per day. The Eight Feet seam of 6ft 6ins thick was producing 700 tons per day. The Brooch Coal of 3ft 4ins thick produced 270 tons per day and the Lower Deep seam of 3ft 7ins thick produced 300 tons per day. Total output per day at this period was 1,620 tons.

The colliery closed on 31 January 1969 because of the difficult geological conditions.

Holly Bank Colliery, Pit No.15, showing the large wooden head frame with its steam winding house and the chimney of the boiler plant.

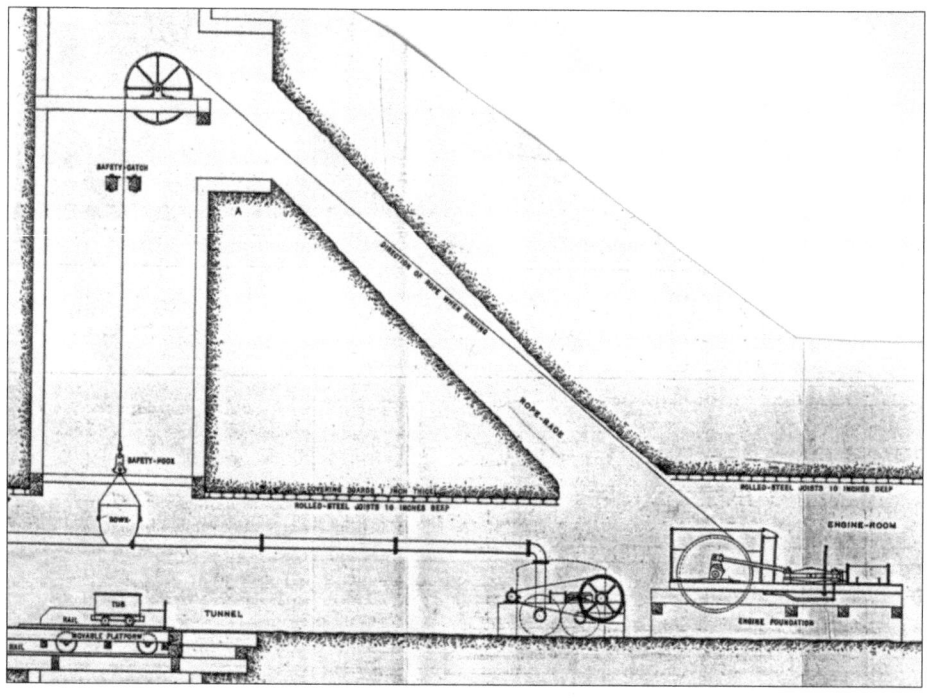

Holly Bank Colliery. This drawing shows the underground winding pulleys and the shaft top with a winding engine driven by compressed air. 1910.

The sinking of the first Hilton shaft in 1924 with the steam plant on the left. Steam was used during the sinking to drive the plant. Then a permanent electric plant was erected to operate the future colliery and the steam plant removed. The foreground appears to have been levelled for the railway sidings.

A group of surface workers at Hilton Main with a coal loading plant in the background. Note several of them hold 'Spraggs' (lengths of wood put into the spokes of wagons to stop them).

A clean group of workers probably joiners pose beside a brand new 10-ton coal wagon. Was this the first wagon built in the colliery workshops?

Locomotive *Holly Bank No. 1*. An example of the products of Messrs Hudswell Clark & Co. supplied to the colliery in 1893.

Downcast shaft top buildings and headgear, the large winding house on the left being one of the few buildings still existing on the site. Most of the site has become the Hilton Park truck stop, but the winding house for a period was part of Squire Transport's depot and was re-clad in grey steel sheets. In the last few years the building has been empty.

Screening plant with railway wagons and the colliery tip in the background about 1960. In the left foreground stands a winch for moving wagons about the sidings.

Cable drums and timber for underground supports stacked in the pit yard. Overhead are the conveyor systems for delivering coal to screening plant or railway wagons. This is believed to have been the 1960s.

Pit yard of about 1960 with timber for underground use stacked on the left. Even with the introduction of steel-powered roof supports, much timber was still used as supports for the roof. A narrow gauge tramway system for supplying materials to the shafts is seen on the right.

Hilton Main Colliery from the railway with the coal screening and loading plant in the foreground. Some of the flat wagons on the right are carrying three detachable boxes of coal each. This system much used on Cannock Chase was designed to reduce breakages of the lump coal.

Hilton Main Colliery, showing the coal merchant's horse and cart for deliveries to Essington and area. These carts were designed to tip so that coal could be easily left outside the house for the family to shovel into the yard.

A group of surface workers outside part of the screens, in about 1920.

Both headgears and buildings with a conical tip in the background.

Downcast shaft headgear with the brickworks in the background, 1936.

Inside the Main Winding House with an air compressor in the foreground and switch gear to the right.

Within the huge Power House was located the Main Winding engine of the colliery. In 1924 it was said to be one of the largest in the country.

Aerial view of the Hilton Main Colliery in the 1920s.

Hilton Main Colliery in the 1930s with both shafts sunk and a large brickworks on the right.

Hilton Main rescue team with breathing sets and their equipment laid out for inspection. By the 1920s every large mining concern was required to organise and train a rescue team for emergencies within the colliery. For major accidents they would be assisted by teams from the Mines Rescue Station set up in every coalfield and trained to a high standard for the work.

The directors of the Holly Bank Colliery Co. Ltd. They were: S.J. Lloyd, Sir William Middlebrook, Col W.E. Harrison, John C. Forrest, and in the background Norman Forrest. The father and son, John C. and Norman Forrest, were to manage and develop the Hilton Main Colliery from its beginning. Col Harrison was the owner of the Brownhills Collieries.

Underground roadway junction after reconstruction in 1955 to ease the gradients and straighten out the curves in order to allow Diesel locomotives to operate underground. At the same time larger mine cars were introduced to replace the small tubs. A fine example of survey work is depicted in the left-hand tunnel.

Above: Side view of the Downcast shaft top with buildings conveyors and two chimneys, c.1960.

Left: Last shift up from the workings of the Hilton Main Colliery on 31 January 1969.

four

New British Iron Co.

The British Iron Co. was formed by a group of London financiers to buy and operate ironworks and mines across the country. They purchased an ironworks at Abersychan in Gwent and then purchased a further ironworks at Ruabon in North Wales. They also began negotiations with John Attwood for the purchase of the Corngreaves Ironworks on the South Staffordshire Coalfield. Eventually in 1826 they purchased the works and commenced iron making from the raw materials found under the estate. However, they also discovered that the value of the works and businesses had been overstated and commenced an action in the Courts for the recover of a considerable portion of the £500,000 purchase money. After many years of litigation, then counter action the case was taken to the House of Lords and a verdict was given against the British Iron Co. This ruined the company who went into liquidation in 1843, but reformed as the New British Iron Co. With huge backing from London finance houses, they went on to rebuild the ironworks with six blast furnaces, develop the production of steel, also sank and operated up to six collieries on the Corngreaves estate.

They were a very forward looking company, building schools and housing for their workers, who appear to have enjoyed a better standard of living than most Black Country miners. They pioneered the use of new methods of mining the Thick Coal and introduced new equipment into their mines. It is no accident that the only remaining example of a Guibal mine ventilating fan-house should stand among the buildings of one of their collieries, built at a period when most of Black Country collieries did not ventilate the workings. The Corngreaves Ironworks were unique in another way in not being built on the edge of the canal system, being several miles away. To overcome this logistical problem a narrow gauge steam locomotive worked railway was constructed from the works to the canal and also to the collieries. By means of this complex railway coal and ironstone was moved to the furnaces or the canal basins and manufactured iron goods were shipped to various parts of the world.

Having become one of the largest well-known concerns in the country, they decided in 1887 during one of the worst depressions in the iron trade to go into voluntary liquidation. Several attempts to sell the iron and steel works as a going concern were tried but failed until they were sold in lots during the 1890s. The collieries were put up for sale in May 1896 but could only raise a bid of £1,500 so were withdrawn. At a later sale most of the collieries were sold to Robert Fellows who had already brought the remains of the ironworks. New Hawne and Timbertree Collieries was sold to Shelagh Garratt & Son and continued producing coal and fireclay until the Miners Strike of 1921 prevented the raising of coal for pumping. Because of the stoppage the workings were flooded and the collieries were abandoned. The buildings of the New Hawne Colliery were converted into a Council yard and preserved in this form for many years. They have now been scheduled as the finest remaining group of colliery buildings in the Black Country.

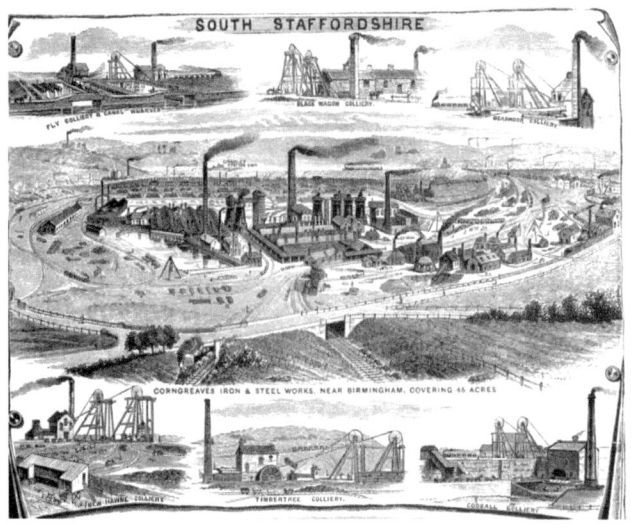

Corngreaves Ironworks, a pen and ink drawing of about 1880, with the two groups of blast furnaces making a total of six furnaces. Much of the ironworks plant and buildings are shown along with collieries and in the foreground a brick-making plant. The railway system in the foreground was from the nearby Timbertrees Colliery.

Corngreaves Ironworks seen from the narrow gauge railway incline from Timbertree Colliery. A number of colliery tubs and several wagons stand in the sidings. The railway was of one metre gauge transporting coal and ironstone from the collieries to the ironworks. The end product in terms of pig iron and castings were then transported to the canal at Fly Colliery and sent all over the world.

Corngreaves Hall today. Formerly the main local offices of the New British Iron Co. The Hall was the subject of a restoration project in 1990. Since then buildings have been left to decay and the vandals.

New Hawne Colliery about 1900 with steam rising from the two plants of the colliery. Near the chimney is the winding house with the Guibal fan house hidden to the right. The shafts on the left were sunk in 1864 to a depth of 268 yards to work the Thick Coal, producing 400 tons of coal per day.

New Hawne Colliery, a group of colliers in front of the winding house. This probably was the total work force at the time. The winding house has two date stones marked 'N B I Co.' and '1865'. Coal and ironstone were produced and screened for dirt before loading into tubs and hauled by rope over a timber viaduct across the valley to feed the furnaces at the Corngreaves Ironworks. New Hawne was one of the first collieries to have a screening plant in the Black Country. The Thick Coal seam here had a shale band in the middle which was removed in the screening plant. The waste was tipped beside the River Stour to create one of the few remaining tips in the Black Country. It was reworked during the Miner's Strike of 1972 for the remaining coal and several thousand tons of coal was extracted. Probably as a result of this disturbance spontaneous combustion set the tips on fire, fortunately for only a short period.

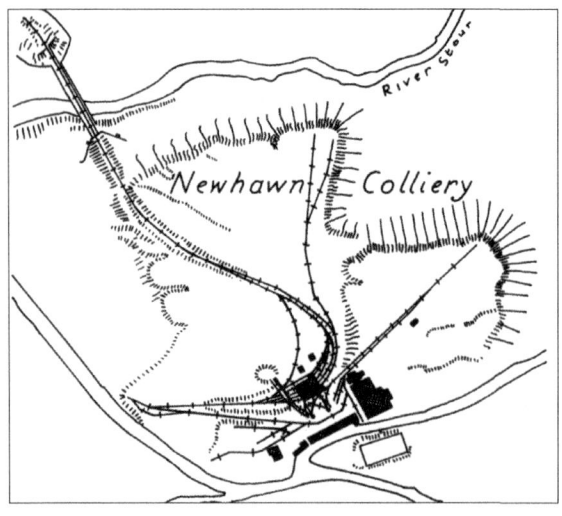

A plan of the New Hawne Colliery in the 1880s. The railway viaduct over the River Stour was originally worked by narrow gauge locomotives. It proved to be too lightly built and swayed in the wind. A rope haulage system was installed and the tubs of coal and ironstone were hauled to sidings near the Corngreaves Ironworks. The final part of the journey was locomotive hauled.

New Hawne Colliery today. The winding house was built in 1865 for a twin-cylinder horizontal steam winding engine built at Corngreaves by the New British Iron Co. The large iron pipe in the side of the building was the exhaust from the steam winder.

A brick-built Guibal fan house built about 1875 to ventilate the workings of New Hawne Colliery. The building is the most complete Guibal fan house in the Midlands. It housed a 21ft diameter fan driven by an 18in horizontal steam engine.

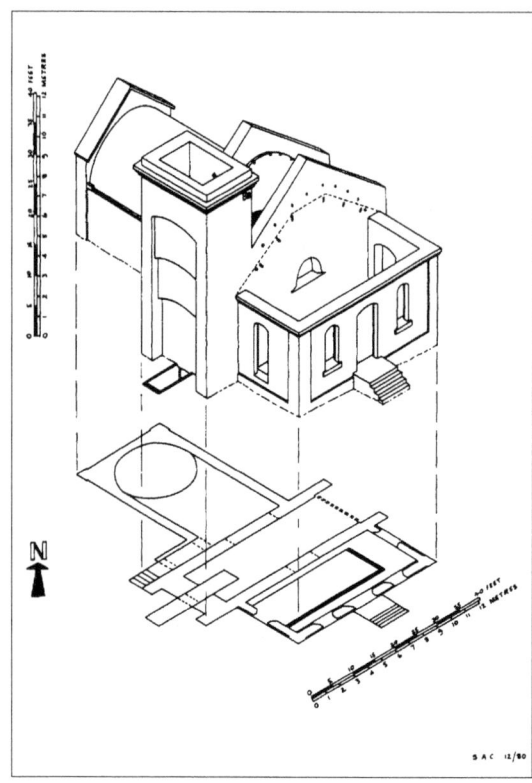

Left: A drawing of a typical Guibal mine ventilating fan house similar to the existing example at New Hawne Colliery.

Below: New Hawne today. A row of workshops with a stables underneath the first floor colliery offices. The offices were added to the single-storey range of workshops in 1895 by Shelagh Garrett & Co. The manager at the time was W.H. Chapman with his brother John as under-manager.

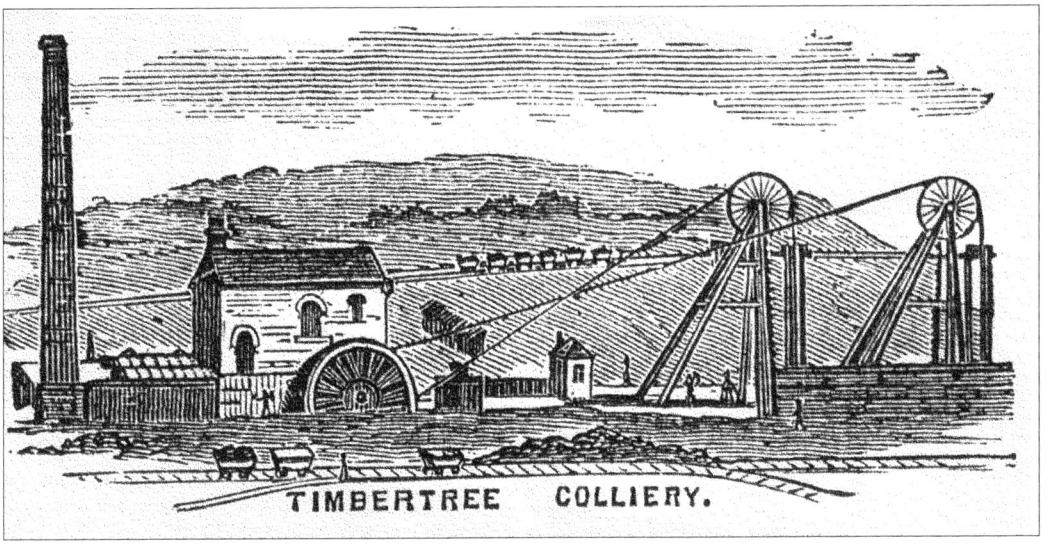

Above: A line drawing of Timbertrees Colliery, sunk in around 1845 by the New British Iron Co. to a depth of 268 yards. Because of being placed at a height above the Comgreaves railway system, a rope-worked, self acting incline was constructed to lower wagons of coal into the yard at Corngreaves. The colliery was closed in March 1915.

Below: Line drawing of Bearmoor Colliery. Sunk before 1842 to a depth of 412ft by the New British Iron Co. The colliery in the six months to June 1842 produced 11,470 tons of coal.

Line drawing of the Black Wagon Colliery. Sunk by the New British Iron Co. before 1850 as part of their plan to extract the Thick Coal from under their Corngreaves estate. The colliery was sold in 1897 to the Corngreaves Furnace Co. who continued extracting coal until taken over by Robert Fellows & Co. in 1902. Closure came in 1926 with the exhaustion of the coal and ironstone seams.

Line drawing of the Codsall Colliery. Working by 1849 and probably sunk by the New British Iron Co. The shafts were 693ft deep to the Thick Coal seam. Following the collapse of the New British Iron Co. the colliery was found to be almost devoid of coal and closed in 1895.

Line drawing of the Fly Colliery on the Dudley No.2 Canal. Sunk in the 1840s the colliery was closed in 1914 but some coal extraction continued until 1926. Pig iron and castings from the Corngreaves Ironworks were transported here for loading into canal boats for transport around the world.

Mabel, an 0-4-0 tank locomotive built in 1899 by Bagnall's of Stafford for the Corngreaves Furnace Co. The Corngreaves Iron and Steel Works were unique in the Black Country by not being built on the side of a canal. Therefore tram-roads were developed from an early date to become narrow gauge railways operating their own fleets of wagons and locomotives.

The New Side blast furnaces of the Corngreaves Ironworks about 1900. Coal from the companies' collieries was converted into coke to fire the furnaces. The hot iron was run into the pig beds in the foreground and allowed to solidify before being sent away. Hot iron was also transport to the steel-making plant in the late 1880s as steel replaced iron for many products. The furnaces survived the collapse of the New British Iron Co. in 1887, were sold to the Corngreaves Furnace Co. in 1894 and to Robert Fellows in 1902. But they were demolished in 1912.

five

Cannock Chase Collieries

Walsall Wood Colliery

The Walsall Wood Colliery Co. was registered on 15 November 1875, while the sinking of two shafts had commenced earlier in the year. Both shafts were completed to the Deep Coal seam at a depth of 576 yards in 1876 and coal production commenced. The colliery during the 1890s was the site of one of the first attempts in this country to apply the German Koepe system of winding with the ropes winding over a pulley instead of onto a drum. Much used at modern collieries this example suffered from rope slip and was eventually replaced. Another early application was the use of steel props and bars in the workings reducing the amount of timber used. Yet up to 1959 the last furnace ventilation system in use at a British colliery was at Walsall Wood. The colliery was to be operated by the National Coal Board from 1947, being extensively modernised in the 1950s. Closure came on 30 October 1964 because of the exhaustion of the coal seams.

Walsall Wood Colliery. Downcast shaft top about 1890 with the heavy timber head frame over the shaft. To the left stands a heavy timber frame with a wheel on the top suggesting the use underground of a rope driven haulage system. The engine house supplying the steam power would be placed on the surface. The shaft was 15ft diameter and 576 yards depth to the Deep Coal seam.

Walsall Wood Colliery, showing the top of the Upcast or ventilation shaft in the 1950s. Under the National Coal Board, the colliery was extensively modernised with a steel headgear over the shaft. Also the steam winder was replaced by an electric winder and the boiler plant demolished. On the far right is the steel exhaust chimney of a ventilating fan, having replaced the last working furnace ventilation system in the country. It seems hard to believe that a fire at the bottom of the Upcast shaft was still providing the ventilation of this colliery until the late 1950s.

Walsall Wood Colliery, showing the valves and cylinder of a Thornewill & Warham steam winder, all highly polished. The cylinders were 42in by 72in stroke with a winding drum of 20ft increasing to 21ft diameter. The engine could lift the cage through the 550 yards depth shaft in 44 seconds.

Walsall Wood Colliery, showing the front of the Thornewill & Warham of Burton-on-Trent built winder. Note the driver stands to operate the levers. These photographs suggest that the engine has recently been rebuilt with new cylinders, probably about 1910.

Walsall Wood, underground in the hauler house with another pioneering plant. This petroleum engine was used to operate an endless rope haulage in the workings. It was the cause of an explosion in 1890 resulting in the death of an engineer and was removed following complaints from the Mines Inspector. A series of photographs featuring the bearded fireman on the right in colliery workings can be identified as having been taken at Walsall Wood from this picture.

THE
WALSALL WOOD COLLIERY CO. LTD.

Best Shallow, Deep and Yard Coal,
BEST QUALITY HOUSE COALS AT THE LAND SALE WHARF.

Also Makers of Blue and Red Bricks, Roofing Tiles, Drain Pipes, Fire-Bricks, Quarries, &c.

PRICES ON APPLICATION.

An advertisement used in the 1930s to display the wares of the colliery company. Many of the collieries developed brickworks or produced tiles and sometimes pipes from high quality clays. A stamp was applied to the bricks with the colliery name; many of these bricks are still to be found about the area.

Walsall Wood Colliery. An early example of an electric bar coal cutter was developed at the Pelsall Collieries before they were taken over by the Walsall Wood Colliery Co. in 1903. The cutter was employed in the workings of Walsall Wood and helped to reduce the cost of mining coal. Known as the 'Simplex' machine, it was developed by Messrs Peake and English, engineers to the Pelsall Colliery.

Walsall Wood Colliery. The Bar Cutter was fitted with wheels which could be raised or lower to suit the height of cutting in the coal seam. Often a shale band would be cut out, allowing the upper coal to fall and be loaded into tubs.

The interior of the Lamp Cabin at Walsall Wood Colliery. During the 1920s the colliery used the Protector safety lamp which was hired from the makers. The boy would be employed to clean and refill the lamps with colsa oil ready for the next shift. Someone would have checked the lamps for damage and attempts to open them. Because of the poor light they gave, colliers tried to take the tops off. This proved to be a source of several explosions so locks with lead rivets were fitted to prevent opening.

Walsall Wood Colliery. The bearded Deputy is examining the effects of the strata above weighing onto the workings. Note the state of the once straight steel girders bent into different shapes by the heavy weight. This gate road would have to be re-cut on several occasions until the weight moved forwards with the advancing coal face.

Right: Walsall Wood Colliery. Colliers working probably in the Deep Coal seam with plenty of height and space.

Below: Walsall Wood Colliery, showing colliers at lunch in the workings. The drinks were often of cold tea. Sometimes a mouse would appear at meal breaks and expect to be fed.

Walsall Wood Colliery. The colliers are undercutting the coal prior to it being dropped and loaded into tubs. They have set several timber props to prevent coal falling on them during the work. The bearded fireman with a Davy safety lamp stands to the left.

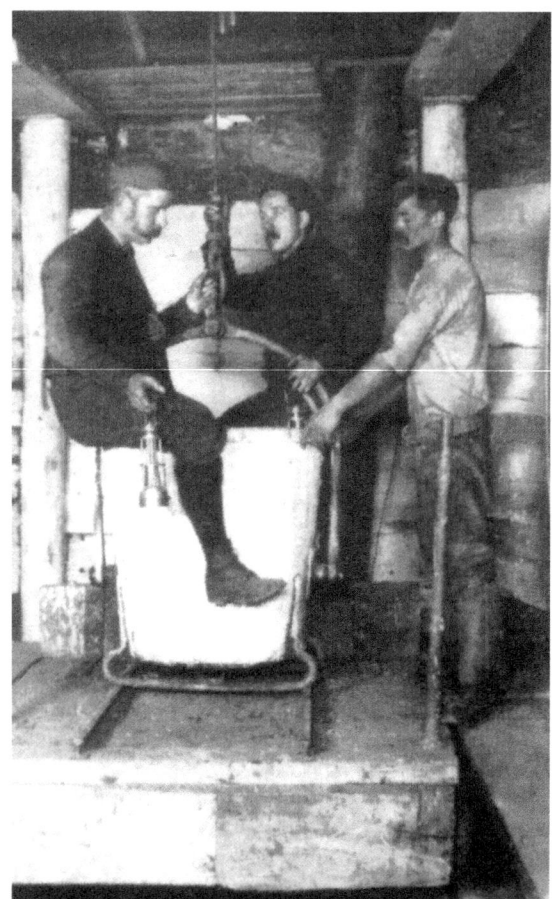

Walsall Wood Colliery. This was how to ride the Bowk in the Upcast shaft – one leg in and one out so that as the Bowk went down the shaft, the outer leg could be used to push away from the shaft sides. The Bowk stands on the sliding door closed over the shaft to prevent air leaks. An early example of a safety lamp is shown here being carried by the colliery manager.

Walsall Wood Colliery. The colliery rescue team displaying their rescue equipment with their instructors. They are all wearing early helmets and two have cap lamps, suggesting a date in the 1940s.

Walsall Wood Canal basin with tubs being pushed over to tip directly into the boat. Tubs could be horse hauled direct from the Downcast shaft to the basin and tipped to await shovelling into the boats.

Walsall Wood Colliery. This aerial view of the colliery was probably taken about 1960 after the replacing of the timber head frames by tall steel frames in the 1950s. Empty tubs are lined up ready to be sent underground at the Downcast shaft.

Pelsall Hall, 1872. This drawing is from a national newspaper and shows the scenes on the pit top during the disaster. Mr Ness, of the Walsall Wood Colliery nearby, was greatly involved in the rescue work. Quickly on the scene were two traction engines that were used to pump large quantities of water over the following five days.

Pelsall Hall Colliery

In the workings of this colliery operated by Messrs Morgan and Starkey, a heading was being driven during November 1872 towards old water-filled workings. The heading unexpectedly cut into the old workings releasing a vast quantity of water into the colliery. Of the thirty-three colliers underground, eleven got out while the twenty-two who were trapped were lost. Pumps from neighbouring and distant collieries were brought to the pit and after five days and nights of work the water was removed and the workings entered. The colliery was to reopen and worked until 1890.

Brereton Colliery

Early coal mining commenced about 1791 when shafts were sunk on Brereton Common. Mining gradually migrated east until between 1872 and 1883 the Brick Kiln pit was developed by the Earl of Shrewbury's company. A new shaft of 15ft diameter was sunk to a depth of 450ft and a new Thornewill and Warham steam winder erected.

Colliers leaving Brereton Colliery in 1926 during a strike. They were being escorted by a car full of police.

Brereton Colliery. Police with motorbikes and side cars wait to escort buses of strike breakers, probably during 1926.

Brereton Colliery pit top from the sidings during heavy snow in February 1940.

Brereton Colliery. Clearing the heavy fall of snow from round the coal wagons ready to bring in the locomotive to take them away.

Brereton Colliery. A side view of the concrete head frame and the snow bound loaded coal wagons. February 1940.

Brereton Colliery, showing a group of ancient six-wheeled coaches standing in the sidings used for transporting the colliers to and from work. Such colliers' transport was usually called 'Paddy Mails'.

Valley Colliery

This was sunk in 1874–75 by the Cannock & Rugeley Colliery Co. to work the Deep and Shallow coal seams. By 1889 all the coal mined at the Valley Colliery was drawn at the nearby Cannock Wood Colliery. During 1922 new steel head frames replaced the old timber frames which were found to be rotten. Valley was under the National Coal Board developed as a training school for the local collieries and since closure has become the Heritage Centre for the area.

Valley Colliery before 1922 with wooden head frames and the winding houses placed between the shafts. Near the left-hand shaft (Upcast) stands a fan house built in 1875 to house a 40ft-diameter Guibal fan.

A similar view taken after the construction of the steel head frames in 1922. Much timber for props to support the roof underground stands in the foreground.

Always an occasion for photographs, the felling of a chimney at the Valley Colliery probably in the 1950s when steam winding was replaced by electric.

An aerial view taken in 1938 to show the new coal screening and washery plant on the left-hand side of the picture. Also looking new is the airlock round the top of the ventilating (Upcast) shaft to the right of the tall single chimney. The steel head frames over the two winding shafts are seen near the tall chimney. The nearest head frame is linked into the new screening plant allowing coal to be conveyed quickly to be cleaned.

Cannock Wood Colliery

Sunk in 1864–65 to a depth of 200 yards by the Cannock & Rugeley Colliery Co., the colliery was one of the success stories of Cannock Chase mining, eventually closing in 1973.

During 1874 a third shaft of 16ft diameter was sunk at Cannock Wood Colliery and a 40ft-diameter Guibal mine ventilating fan was erected to replace the old furnace method. Reduced to standby, the fan was still complete with its steam engine at closure in 1973. Manufactured by Black Hawthorn of Gateshead, the steam engine was dismantled and taken to the Beamish Open Air Museum at Stanley, County Durham, where – fully restored – it is intended to drive their colliery fan.

Cannock Wood Colliery. Today a fine range of colliery workshops still stand on the site as part of the Cannock Wood Industrial estate. Probably constructed in the 1860s, these buildings have found a new life as a Joinery Works.

Pool Lane Colliery

Above: Sunk in the 1930s by Messrs J. & B. Cox, this small and primitive colliery was later operated by J. Williams & Son. With two timber head frames and a simple winding plant, coal was mined and sold to householders.

Left: Colliers with a Bowk on top of a shaft probably at the Pool Lane Colliery. The collier in the left foreground wears a jacket cut away at the front but long at the back to protect him from water when bending down. Usually made of a thick woollen material tightly woven to make them waterproof, these jackets were much used by shaft sinkers. Hanging on a post in front of this collier is a Duck lamp with a long spout which held the oiled wick to provide the light.

Lea Hall Colliery Rugeley

The first new colliery to be planned and sunk by the National Coal Board in the period 1954 to 1960, it was designed to produce over a million tons of coal annually. This magic figure was achieved in 1963, rising to 1,789,681 tons in 1975, a record for a single British colliery. Designed to produce only power station coal delivering directly into the adjacent Trent Valley Power Station, the colliery maintained a high output until closed in December 1990.

Construction of the electric winding house in the 1950s with a small sinking head frame hidden under the permanent head frame. The shaft was large enough to take four separate cages working within its confines.

Lea Hall Colliery in April 1991 with the two huge head frames towering over the buildings. The use of up to four cages per shaft made the high output of over 1,000,000 tons of coal annually possible.

Cannock and Leacroft Colliery

The Cannock & Leacroft Colliery Co. were registered in 1871 with a capital of £50,000. Between 1874 and 1877 they sank two shafts to a depth of 389 yards to work both the Shallow and Deep coal seams of Cannock Chase.

Left: The front of the winding house and a superb lattice steel head frame over the Downcast shaft at the Leacroft Colliery. Sunk in 1874 to a depth of 407 yards, the colliery was merged with Mid-Cannock in 1954 and all coal winding ceased. During 1959 a pumping plant was erected using the Downcast shaft to assist the pumps at other collieries. With further colliery closures the pumping plant ceased operations on 20 June 1968 and the shaft was filled during July 1968.

The Norton Cannock Colliery Co. was registered in 1874 with a capital of £50,000 and the sinking of two shafts was commenced in 1875. Coal was being produced by 1876 but the company was soon in liquidation. A new company known as the Norton Cannock Coal Co. was formed in 1877 and coal production was commenced again. The colliery was to operate until the coal seams were exhausted by 1910. The plant on the site was for sale in October 1910.

West Cannock Colliery. Promoted by William Molyneaux, the first shaft was sunk from 1869 to 1871 to cut the Deep Coal seam at a depth of 310 yards. During the 1890s an output of 1,800 tons of coal could be raised in eight hours. Placed on a north to south line were three shafts, No.4 was 124 yards deep to the Main Hard Coal seam, the Middle Shaft served to ventilate all the mines and No.1 shaft was 10ft diameter and 310 yards deep.

Opposite below: A snowy scene at Wimblebury Colliery about 1960. Sunk from 1873 onwards, the colliery was the scene of a major underground fire in July 1880. As a result the Cannock & Wimblebury Colliery Co. failed in November 1880 and a sale took place in 1881. A New Cannock & Wimblebury Colliery Co. was formed in May 1881 and liquated in 1887. The colliery including workings was sold in 1887 to the adjacent Cannock & Rugeley Colliery Co. who merged the mine with their Valley Colliery. In 1947 the output was 181,682 tons with 900 men, by 1962 output was 208,233 tons from 776 men. The colliery was finally closed in December 1962.

Above: West Cannock Colliery with on the left the ventilating shaft wreathed in warm air rising out of the mine. With steam rising from the pipe to the left of the chimney, it suggests the winder was working at the time the photograph was taken. In the background the rolling hills of Cannock Chase are seen.

Grove Colliery

From 1849 William Harrison had operated a series of collieries under the name of the Brownhills Colliery Co. With a lease from Phineas Hussey, the company were able to sink in 1869 the Grove Colliery shafts near Norton Canes. Situated beside the Cannock Extension Canal, a bridge was built to extend the railway over the canal into the colliery yard. Closed by the National Coal Board in January 1952 after merging with the Wyrley No.3 Colliery in November 1950. The coal screening plant was used to process coal from the Wyrley No.3 colliery about a mile away using a tramway with rope haulage.

Above: Grove Colliery. Lines of loaded coal wagons await collection while in the foreground a horse drawn cart is loaded by hand for a local sale. In the background tubs of coal from the shaft are being hauled up a ramp into the screening plant. The leisurely method of loading the cart by hand contrasts with the more rapid loading of railway wagons at the screens.

Opposite below: Number 3 Plant consisted of two shafts 10 yards apart and sunk, 373 yards to the Deep Coal seam. In 1893 it had recently been reopened and new plant and a large screening plant erected. Over the Downcast shaft was a tall complex timber head frame operated by a single-cylinder winding engine working two cages with one tub per deck.

Grove Colliery. A timber head frame dominates the screening plant over the railway sidings with loaded coal wagons standing in the sidings.

Brownhills Colliery. A group of carpenters with their tools stand beside their handiwork outside the workshops of the company. Having manufactured and painted to a high state this coal wagon, it would be taken to one of the collieries and would then have 10 tons of coal dropped in it!

Hednesford Colliery

Sunk by Messrs Webb and Poxon about 1850 to a depth of 48ft to work the Seven Feet coal. Taken over by Messrs Piggott & Co. in the 1850s, they deepened the shafts to work the Brooch and Second Coal seams in 1859. Francis Piggott also worked the Gubbin ironstone seam by means of a quarry which was much visited by local geological societies during the second half of the nineteenth century. Sold in 1870 to the Cannock Chase Colliery Co., Hednesford Colliery became Cannock Chase Collieries Nos 9 and 10.

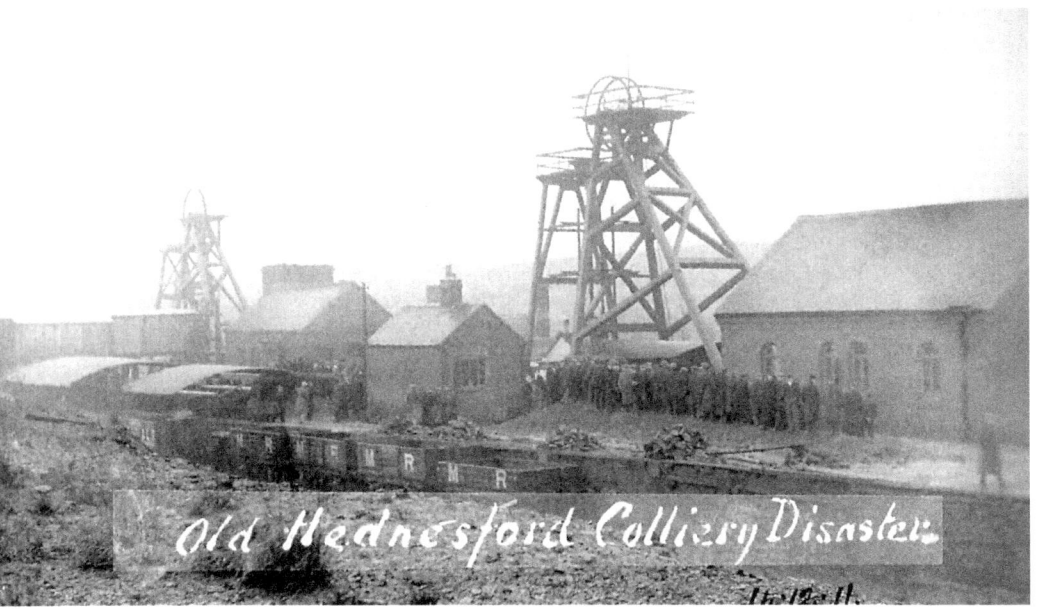

Hednesford Colliery. It is an unfortunate fact that a colliery accident involving some loss of life was also a time when some of the best photographs of collieries were taken. An explosion of fire-damp took place during January 1871 when Joel Chilton was killed and two other colliers were burnt. The disaster mentioned in this picture took place on 14 December 1911 and resulted in the deaths of five colliers and twenty horses.

Left: S.F. Sopwith was manager of the Cannock Chase Collieries in the 1920s and developed an early form of pit prop. Later he improved his prop by using a steel girder in place of timber. A classic straight coal face with two rows of Sopwith props over a conveyor taking coal to the gate-roads to be loaded into tubs.

Below: Underground in the Cannock Chase No.3 Colliery. The collier on the right is operating one of Sopwith's hydraulic props. On the left stands a pony wearing a leather head protection for working in low places. The metal cones were to protect the pony's eyes from sharp objects in the dark levels of the colliery.

Cannock Chase Colliery

Commenced in 1854 by John McClean and Richard Croft Chawner, the Cannock Chase Colliery Co. rose to operate ten collieries on the Chase and became the major coal producer of the area. Of these mines, No.3 pit was sunk about 1862 to a depth of 165 yards to work the Deep Coal seam. In 1947 these collieries were taken over by the National Coal Broad, the No.3 pits eventually closed in 1961. By this date very little coal remained to be extracted.

Right: Two colliers with hand drill in the workings of the Cannock Chase No.3 Pit. The collier on the left holds an early Clanny flame safety lamp. This form of lamp incorporated a glass tube round the flame to provide more light than the normal wire gorse-covered safety lamps.

Below: Mid-Cannock Colliery in the 1950s with the downcast shaft on the left and the Upcast with its airlock in the head frame on the right. The rural nature of the colliery at that time is shown by the stooks of corn in the nearby fields.

Mid Cannock Colliery

Sunk from 1873–75 to work the Deep and Shallow coal seams over about 40 acres, the colliery was closed in 1880. A sale of the plant took place on 15 December 1881 at the Queens Hotel in Birmingham. The shafts were filled and left until reopened in 1913 by William Harrison Limited. In 1947 the output was 217,258 tons from 845 men. It closed in December 1967, when the output was 171,010 tons with 871 men employed. The shafts of 270 yards' depth were used as a pumping plant to keep other local collieries working until filled with washery dirt from Littleton Colliery in October 1968.

East Cannock Colliery

On Wednesday March 25th 1874 the first sod of new pair of shafts to be sunk by the East Cannock Co. on the Hemlocks Farm was turned by Miss Silvester of Stafford in the presence of a goodly number of ladies and gentlemen connected with the enterprise. Afterwards the party dined at the Anglesey Hotel Hednesford, to jointly celebrate the event and also to express their gratification at finding the Deep and Shallow seams already at the 'Amy' sinking.
Mining Journal of 28 March 1874

The sinking of two shafts was commenced in 1871 with the first coal wound in 1874. The company was liquated in 1880 and the colliery purchased in October by a new company. It was noted that the colliery had cost around £150,000 to create but was sold to the new concern for £20,000. They were to operate the colliery until Nationalisation in 1947 and went into voluntary liquidation on 5 September 1951. In 1947 the output was 211,018 tons. The colliery was closed in May 1957 while the last year's output was 186,193 tons.

The tip from East Cannock Colliery towers over a cottage at Rumer Hill, Hednesford. The cottage appears to be occupied; it must have been a dusty life, especially on washing day.

The Downcast shaft with timber head frame and winder house probably about 1900. The shafts were sunk to a depth of 368 yards to work the Deep Coal seam.

East Cannock Colliery in 1957 with many of the alterations undertaken by the National Coal Broad after 1947. Most notable of these was the replacement of the timber head frames by steel frames complete with large frames to lift the pulley wheels on the top.

One of the effects of coal mining under buildings was the subsidence that often occurred. This farmhouse appears to lean to the left and has been propped up with timber to prevent further movement. Mining leases in the area had clauses such as 'no liability for surface damage'. In the background the East Cannock Colliery is seen.

Essington Wood. Two colliers with a hand winch working shallow seams of coal in the 1920s. Mining in the area commenced about 1790 on a series of shallow coal seams and has continued almost to the present day with open cast operations by contractors into the 1990s.

While Cannock Chase developed a less extensive a network of canals than the Black Country, most early collieries were sunk close to canals for transport. By the 1860s the construction of railways to serve the coalfield was undertaken and both forms of transport were to move huge tonnages of coal to many different places, including London. The canal boats seen here at Hednesford, loaded and empty, are probably about 1905.

Sinking of the Littleton Colliery shafts in 1898. The temporary building houses the steam plant to raise and lower the Bowk or bucket in the shaft. The building in the head frame held small steam engines for lowering the pumps as the shaft was sunk. Three steam pumps were used in the shaft; each was capable of lifting 1,000 gallons per minute. The capacity of these pumps was to be tested by the huge volumes of water encountered in sinking these shafts.

Littleton Colliery

This was commenced as the Cannock & Huntingdon Colliery Co. and registered on 3 May 1873. Being aware of the huge amounts of water lodged in the Bunter pebble beds, they attempted to sink the two shafts using the Belgian Kind Chauldon method of boring followed by inserting iron tubbing. After making the iron tubbing watertight the shaft would be sunk by hand. Having got to a depth of 438ft in 1880, water broke in and flooded the shafts. After attempts to raise new capital to continue the sinking, the company was wound up in 1884. Recommenced in April 1897 by Lord Hatherton, the landowner, the No.2 shaft was sunk to a depth of 1,644ft on 17 February 1899. Attempts to recover the No.1 shaft were abandoned on 3 May 1900 and a new shaft was commenced. This sinking after dealing with huge quantities of water was completed at a depth of 1,662ft on 22 November 1902. As the Littleton Collieries Ltd the colliery was to be one of the most successful on Cannock Chase, lasting into National Coal Board days before closing as the last working colliery on the Cannock Chase coalfield in 1994.

Front of the sinking head frame with a Bowk of spoil from the shaft bottom being tipped into a wagon on the shaft top. The collection of oilskins on the hovel on the left illustrates the conditions underground.

Littleton Colliery. This scene is from about 1920 with the three head frames, all with pulley wheels and a smoking chimney in the background. It could easily be a Sunday view.

Littleton Colliery. The tall steel headgear over the Upcast shaft and the 'Evasse' chimney of the mine ventilating fan on the right. The photograph was taken in 1994, after the colliery had closed.

Above: Littleton Colliery. The colliery offices were built about 1900; they were still in use and well maintained in 1994. It was hoped to preserve them but unfortunately they were demolished.

Left: Littleton Colliery. The Downcast shaft top and steel head frame in 1994. The shaft had been equipped with skips to improve the output in the 1960s making the magic figure of 1,000,000 tons per year output a regular feature of the latter years of the colliery.

Opposite above: Littleton Colliery. Extensively modernised by the National Coal Board in the 1960s including the driving of new straight and level tunnels for locomotives. At the same time mine cars were introduced to speed the delivery of coal to the bunkers at the shaft bottom. From these the skips in the shaft were loaded.

Opposite below: Littleton Colliery, showing the end of a conveyor system on a gate-road from the workings to the pit bottom. Conveyors delivered coal to the shaft bunkers, gradually replacing the locomotives at some collieries.

Littleton Colliery. Underground shaft being sunk in August 1974 to create a coal bunker with a capacity of 600 tons of coal. Placed near the shaft the bunker was designed to speed the flow of coal to the skips running in the shaft.

Littleton Colliery. Putting into place a concrete section of the shaft lining in the coal bunker during August 1974. The dry conditions should be noted. Most shaft sinking was done in water which cascaded from the sides or rose from the floor, with the Sinkers employed being clad in oilskins.

Conduct Colliery

Sunk about 1856 by Jerome Clapp Jerome, the colliery suffered from water problems, spending the company's money. Liquidated in May 1860, the royalty was taken over by James and Charles Holcroft of Bilston, who sank new shafts in 1866 and created the Conduct Colliery Co. Taken over in 1931 by the Littleton Colliery Co., the colliery was closed in 1933.

On the left has been sunk a large diameter shaft with a tall engine house probably for twin-cylinder vertical winder. This was the main output shaft of the colliery. In the middle stands a fine example of the Guibal mine ventilating fan house wreathed in steam to show it was working. On the right are two separate head frames over small diameter shafts probably the site of the original sinking of the colliery. In the foreground are extensive tips overlooking the railway sidings.

The canal basin at Conduct Colliery with loaded and empty boats in a confused jumble. Note the tubs pushed over on the right-hand side to speed loading of the boats and save on back breaking toil. One is left wondering how the boats got out of this!

Coppice Colliery

Sinking of the shafts was financed by R.W. Hanbury and the colliery was often known as the 'Fair Lady' after his wife who performed the cutting of the first turf. The family retained ownership of the colliery until 1947. At the time the daily output was about 530 tons. Under the National Coal Board the colliery was the first in the area to receive powered cutting and loading machines raising the daily output to 1,100 tons. In 1960 with a work force of 360 men an output of 202,000 tons was produced. Closure came in 1964 because of the exhaustion of the coal seams.

Coppice Colliery in a superbly posed picture with the company's wagons lined up to identify the colliery. It was probably taken in the 1890s when the colliery was newly sunk.

Coppice Colliery with a fine lattice steel head frame in the 1950s, after modernisation. In the 1954 the colliery achieved a national output record of 81cwt per man shift, at the time more than three times the national average.

Coppice Colliery and the mines' rescue team, complete with the van to move them quickly to collieries as required. The van suggests they were expected to travel around Cannock Chase, as was required for major incidents. This was likely around the 1920s.

Nook Colliery

Working from 1879 to 1949, this small colliery was operated by the Nook & Wyrley Collieries Co. Taken over by the National Coal Board in 1947, it was closed in June 1949. This view is from around 1960 when the plant was being operated as a pumping station for other local mines.

Spring Meadow Colliery

Situated at Cheslyn Hay, this small colliery was working in 1945. Coal in the tubs shown was brought out of the Essington drift and up onto the elevated tipping level for screening and loading into lorries.

Spring Meadow Colliery, showing the pit yard with a jumble of tubs and the elevated loading platform in the background. It was taken after the colliery had been abandoned in 1967.

Swan Colliery

Situated at Brownhills, the buildings of this small colliery are shown.

Poplars Opencast

Massive National Coal Board walking dragline working on the shallow coal seams exposed at the Poplars site. Coal was then taken by lorries to the Mid Cannock Colliery washer for processing before despatch. A huge hole was created by the removal of the coal, providing a local waste disposal company with an opportunity to infill and create a hillside.

Other local titles published by The History Press

Made in Walsall
MICHAEL GLASSON

Four themes of ingenuity, adaptability, receptiveness and exceptional craftsmanship run through this book. Michael Glasson provides an illustrated history of industry in Walsall, the roots of which can be traced to long before the Industrial Revolution.

07524 3566 3

Birmingham Canalside Industries
RAY SHILL

The book looks at the many canalside industries of Birmingham and the Black Country, such as iron, coal, gas, electricity, bricks and firebricks, and railway interchange, plus some of the more modern trades. It is accompanied by hundreds of intriguing old photographs.

07524 3262 1

Walsall Leather Industry The World's Saddlers
MICHAEL GLASSON

For nearly two hundred years the Midlands town of Walsall has been a major centre of the leather industry, and has sometimes even been called the saddlery 'capital' of the world. Michael Glasson explores the history of this fascinating industry.

07524 2793 8

The Birmingham Gun Trade
DAVID WILLIAMS

Complemented by over 130 images, this book discusses gunmaking in Birmingham, craft skills, machines and most importantly the effect of changing technoligies on people's jobs, livelihoods and locations.

07524 3237 0

If you are interested in purchasing other books published by The History Press, or in case you have difficulty finding any History Press books in your local bookshop, you can also place orders directly through our website

www.thehistorypress.co.uk